EYE IN THE SKY

SMITHSONIAN HISTORY OF AVIATION SERIES
Von Hardesty and Michael H. Gorn, Series Editors

On December 17, 1903, human flight became a reality when Orville Wright piloted the *Wright Flyer* across a 120-foot course above the sands at Kitty Hawk, North Carolina. That awe-inspiring 12 seconds of manned, powered flight inaugurated a new technology and a new era. The airplane quickly evolved as a means of transportation and a weapon of war. Flying faster, farther, and higher, airplanes soon encircled the globe, dramatically altering human perceptions of time and space. The dream of flight appeared to be without bounds. Having conquered the skies, the heirs to the Wrights eventually orbited the Earth and landed on the Moon.

Aerospace history is punctuated with many triumphs, acts of heroism, and technological achievements. But that same history also showcases technological failures and the devastating impact of aviation technology in modern warfare. As adapted to modern life, the airplane—as with many other important technological breakthroughs—mirrors both the genius, as well as the darker impulses of its creators. For millions, however, commercial aviation provides safe, reliable, and inexpensive travel for business and leisure.

This book series chronicles the development of aerospace technology in all its manifestations and subtlety. International in scope, this scholarly series includes original monographs, biographies, reprints of out-of-print classics, translations, and reference materials. Both civil and military themes are included, along with a systematic study of the cultural impact of the airplane. Together, these diverse titles contribute to our overall understanding of aeronautical technology and its evolution.

EYE IN THE SKY

THE STORY OF THE CORONA SPY SATELLITES

EDITED BY DWAYNE A. DAY,
JOHN M. LOGSDON,
AND BRIAN LATELL

SMITHSONIAN INSTITUTION PRESS
WASHINGTON AND LONDON

Copy Editor: Initial Cap Editorial Services
Production Editors: Ruth Thomson and Robert A. Poarch

Library of Congress Cataloging-in-Publication Data
Eye in the sky : the story of the Corona spy satellites / edited by
Dwayne A. Day, John M. Logsdon, and Brian Latell.
p. cm. — (Smithsonian history of aviation series)
Includes index.
ISBN 1-56098-830-4 (cloth: alk. paper)
1. Project Corona (United States)—History. 2. Space surveillance—United States—History. I. Day, Dwayne A. II. Logsdon, John M., 1937— . III. Latell, Brian. IV. Series.
UG1523.E94 1998
327.12—dc21 97-19238

British Library Cataloguing-in-Publication Data available

A paperback reissue (ISBN 1-56098-773-1) of the original cloth edition

Manufactured in the United States of America
04 03 02 01 00 99 5 4 3 2

♾ The paper used in this publication meets the minimum requirements of the American National Standard for Information Sciences—Permanence of Paper for Printed Library Materials Z39.48-1984.

Dedicated to the CORONA Pioneers

CONTENTS

PART TWO. VOICES OF THE CORONA PIONEERS

ACKNOWLEDGMENTS

An edited book on a long-secret topic such as the first U.S. photoreconnaissance satellite is clearly the product of many more people than the volume's essay authors and editors. First of all, the information required to make this book possible had to be declassified, and thus we want to acknowledge those in the White House, Congress, and the Intelligence Community who have led the movement toward increased openness—particularly Vice-President Al Gore and former Director of Central Intelligence Robert Gates. We wish to thank everyone who provided photographs for possible inclusion, information on topics covered, or comments on the manuscript as it evolved. Many of these individuals are mentioned by name below or in the individual essays, but we have not mentioned everyone who deserves credit for helping to create or improve this book. We owe them all a debt of gratitude.

The chain of events that led to *Eye in the Sky* began with a conversation between one of us, Dwayne A. Day, and Carole Minor, then academic coordinator for the Center for the Study of Intelligence of the Central Intelligence Agency. Carole enthusiastically agreed with Dwayne's suggestion that an already planned conference marking the declassification of the CORONA program and imagery be cohosted by the Space Policy Institute of George Washington University's Elliott School of International Affairs and held on the GW campus. That conference, which took place in May 1995, was a great success, prompting a decision to continue the collaboration by coediting this book. Carole stayed involved with the conference and its follow-up activities, and was a source of invaluable assistance and good advice throughout.

R. Cargill Hall suggested after the conference that its proceedings could form the basis for an interesting and valuable book, and he pointed us to Von Hardesty of the National Air and Space Museum, who is the editor of the Smithsonian

Institution Press series on aerospace history. Von recognized the merit of such a book and he introduced us to Mark Hirsch of the Press, who guided the book proposal and manuscript through the review and acceptance process. We did not always agree with Mark's ideas on an initial hearing, but they have had a strong influence on making the book not just a record of conference proceedings, but primarily a set of original essays on CORONA and strategic intelligence. The two anonymous reviewers of the prospectus for the book also provided ideas that we were happy to incorporate into the book's design.

A number of people provided help in gathering photographs from the volume. They include Sam Welch at the National Archives, and Guy Aceto, who facilitated access to the photo collection of *Air Force* magazine. Dino Brugioni, James Plummer, Chris Pocock, Charles Vick, and A. Roy Burks provided photographs from their cherished personal collections. The Space and Missile Systems History Office at Los Angeles Air Force Base also was very helpful. Dick Wollensak of Itek also provided valuable material.

The National Reconnaissance Office, just emerging from more than three decades of highly classified existence, initially was somewhat skeptical of the cooperation between the Space Policy Institute and the CIA in organizing the 1995 conference and then this book. For an organization whose very existence had been officially denied for so long, acceding the open discussion of former state secrets was not easy. But NRO has been rapidly adjusting to new realities, and has been helpful in providing material for this work. Special mention should be made of former NRO public affairs officer Katherine Schneider and historian Gerald Haines, who led us to useful material and answered questions about the early years of the NRO.

At the Space Policy Institute, David Onkst has served as almost a fourth editor of this book. In particular, he took the lead in putting together the excerpts from the remarks of speakers at the May 1995 conference that comprise Part Two of this book, and he prepared the index. Anne Marie Burns, Garth Henning, and Bridget Ziegelaar also helped along the way.

At the Center for the Study of Intelligence, several people were vital to making the 1995 CORONA conference possible and thereby providing much of the impetus and substance for this book. In July 1993 David Gries, who was the center's director from 1992 to 1994, called the first exploratory meeting to discuss a CORONA conference; he guided the delicate detailed planning for such a conference for more than a year. David Doyle, A. Roy Burks, and Mary Burns provided the glue that held the conference organizing process together. Kevin C. Ruffner was the CIA historian who edited the illuminating volume *CORONA: America's First Satellite Program,* which was distributed at the conference. That book of declassified intelligence documents laid the foundation for this study of CORONA.

For the Smithsonian Institution Press, Therese Boyd provided skillful editing of the manuscript and Ruth Thomson guided the book through the publication process.

The lasting value of this book will be based on the accounts of the CORONA program and the Cold War environment within which it emerged and was implemented that are provided by the essay authors in part 1 of the book and the representatives of the CORONA Pioneers—the informal group of those who worked in secrecy to bring the program to fruition—whose accounts of their part in the effort are included in Part Two. As editors, we have learned from them all. Our thanks are just a small portion of the gratitude owed by U.S. citizens to all the CORONA Pioneers, and to those that followed them throughout the lifetime of the program.

Dwayne A. Day
John M. Logsdon
Brian Latell

ABBREVIATIONS

ABM	antiballistic missile
ARC	Ad Hoc Requirements Committee
ARDC	Air Research and Development Command
ARPA	Advanced Research Projects Agency
CEP	circular error probable
CIA	Central Intelligence Agency
COMOR	Committee on Overhead Reconnaissance
DCI	Director of Central Intelligence
DDE	Dwight D. Eisenhower Library
DoD	Department of Defense
ICBM	intercontinental ballistic missile
IGY	International Geophysical Year
IMC	image motion compensation
IRBM	intermediate-range ballistic missile
JCS	Joint Chiefs of Staff
LC	Library of Congress
NARA	National Archives and Records Administration
NIE	National Intelligence Estimate
NRO	National Reconnaissance Office
PFIAB	President's Foreign Intelligence Advisory Board
SAC	Strategic Air Command
SAM	surface-to-air missile
SIRC	Satellite Intelligence Requirements Committee
SRV	satellite recovery vehicle
TCP	Technological Capabilities Panel
USGS	U.S. Geological Survey
USIB	U.S. Intelligence Board

DWAYNE A. DAY, JOHN M. LOGSDON, AND BRIAN LATELL

INTRODUCTION

> **I wouldn't want to be quoted on this but we've spent 35 or 40 billion dollars on the space program. And if nothing else had come out of it except the knowledge we've gained from space photography, it would be worth 10 times what the whole program has cost. Because tonight we know how many missiles the enemy has and, it turned out, our guesses were way off. We were doing things we didn't need to do. We were building things we didn't need to build. We were harboring fears we didn't need to harbor. Because of satellites, I *know* how many missiles the enemy has.**
>
> President Lyndon Johnson, March 16, 1967[1]

Satellite reconnaissance profoundly affected the Cold War, but it existed in such secrecy that its contributions to international stability during that conflict have been largely unheralded. This book, the first comprehensive work about the early U.S. reconnaissance satellite programs based on authoritative sources, is intended to correct that. Although scholars and popular writers have written about reconnaissance, none have had the access to the official records and personnel that the editors and contributors of this book possessed. The information on these pages is based on formerly top secret records. Many of the photographs have never been published before. Many of the people quoted here are legends within the Intelligence Community but virtually unknown outside it. This is the first time they have spoken in an unclassified format. Consequently, this book is the first full account of the most significant development in intelligence in the twentieth century—the reconnaissance satellite.

Discoverer IV, CORONA Mission 9001, lifting off on June 25, 1959. This was the first rocket to carry a reconnaissance payload. However, the Agena upper stage malfunctioned and the satellite did not enter orbit. (Photo courtesy of *Air Force* magazine)

THE IMPACT OF PEARL HARBOR

The December 1941 surprise attack on Pearl Harbor was one of the most profound events in American history. In military terms, the attack was not terribly successful. Most of the ships "sunk" at Pearl Harbor were raised and repaired within months, and the attack only served to invite the wrath of a sleeping giant. Within less than four years, the United States emerged as the predominant military and economic power on the planet.

Yet Pearl Harbor had an effect far beyond its immediate military impact. It served as a forceful warning of the dangers of a surprise attack. Members of the military and intelligence communities applied what they learned from this lesson in different ways.

For many in the military, the lesson meant to be prepared for all contingencies. The natural inclination for military leaders was to plan for the worst-case scenario. Pearl Harbor meant more aircraft, more tanks, more ships, better security, and peace through strength. It meant that the United States needed forces able to strike back at an adversary hard and immediately. It meant that the United States needed to be so strong that no sane adversary would consider attacking it. Furthermore, this lesson meant that the price of security was measured in material things, such as defense budgets and weapons systems.

For many in the Intelligence Community, even within the intelligence branches of the U.S. military, Pearl Harbor was a warning of the dangers of not knowing what America's potential adversaries were planning and capable of doing. The ability to know what the enemy could and would do did not necessarily mean that the country required large military forces. Accurate intelligence meant that the country could plan better and more economically. Intelligence was a "force multiplier"—it dramatically enhanced the power of existing forces and illustrated how they could be employed more wisely.

Early developments in the Cold War enhanced these differing lessons. The Soviet detonation of an atomic bomb in 1949 was only one of the first harsh warnings. U.S. leaders were not only surprised by the Soviets' rapid development of the atomic bomb (made possible through Soviet acquisition of American bomb plans), but also shocked, albeit to a lesser extent, that they learned of the Soviet bomb only *after* it had been successfully tested. In 1950, North Korean forces, with the approval of Josef Stalin, invaded South Korea and sent another shock through the American military and intelligence systems. In 1953 the Soviets detonated a hydrogen bomb; once again the United States detected this only after the fact. The United States was now faced with an adversary that could attack at any moment with weapons of incredible destructive power.

The military's response was to call for more bombers, more nuclear weapons, and the capability to survive a Soviet first strike and to hit back with a massive retaliatory strike. The Intelligence Community's response was to develop better intelligence on the Soviet Union and to start learning what the Soviets were preparing to do *before* they actually did it.

These two attitudes prevailed during the formative years of the Cold War and collectively shaped the American response to the Soviet threat. It was because of Pearl Harbor, the Soviet development of atomic and hydrogen weapons, and the invasion of Korea that the United States began to seek the means to carry out strategic reconnaissance of the Soviet Union. It was for these reasons that the United States developed the U-2 reconnaissance aircraft, and for these reasons that the United States developed a reconnaissance satellite program known as CORONA. That program is the focus of this book.

CORONA revolutionized intelligence. Whereas in 1949 and 1953 the United States learned of Soviet nuclear weapons tests only after they had occurred, in 1964 the United States learned of the Chinese development of an atomic bomb *before* it had been tested and used this information to blunt the propaganda effect of the Chinese achievement. Satellite reconnaissance enabled the United States to determine that there was no missile gap, and therefore to lower dramatically the number of missiles built to counterbalance the Soviet threat. Satellite reconnaissance also provided a firm basis for arms control, since it removed most of the guesswork from assessments of Soviet strategic capabilities. The supreme value of satellite reconnaissance was that, unlike some human agents, satellite photographs did not lie.

Satellite reconnaissance had a profound effect upon the conduct of the Cold War. Yet, except for Lyndon Johnson's off-the-record but widely quoted remarks in 1967, no American president officially mentioned reconnaissance satellites until Jimmy Carter in 1978. The existence of the super-secret organization responsible for managing the U.S. spy satellite program was not officially admitted until 1992. Everything about reconnaissance satellites was cloaked within a heavy veil of secrecy. That has now changed.

A PROGRAM DECLASSIFIED

In September 1996, the White House issued a new statement of National Space Policy. That statement acknowledged that "the United States conducts satellite photoreconnaissance for peaceful purposes, including intelligence collection and monitoring arms control agreements."[2] This acknowledgment represented the formal lifting of the security veil behind which the U.S. program of photoreconnaissance from orbit had operated since 1958.

An early CORONA payload and reentry vehicle. The reentry vehicle is the gray nose section. The camera's lens cover is not visible in this photo. Early versions of the CORONA payload were small compared to those produced a decade later. (Photo courtesy of A. Roy Burks)

As Albert Wheelon, one of those most involved in creating a continuing program of reconnaissance satellites, says in chapter 2 of this volume,

> When the American government eventually reveals the [full range of] reconnaissance systems developed by this nation, the public will learn of space achievements every bit as impressive as the Apollo Moon landings. One program proceeded in utmost secrecy, the other on national television. One steadied the resolve of the American public; the other steadied the resolve of American presidents.[3]

CORONA

Central to the overall reconnaissance satellite effort was a program known as CORONA. (It is U.S. Intelligence Community practice that the code names of specific programs or projects are always capitalized, whereas other military space programs at that time were usually capitalized, but not always in public documents.) CORONA had its origins in proposals first developed at the RAND Corporation, the civilian "think tank" that supported Air Force strategic planning. As it considered possible security-related and military uses of space in the early 1950s, RAND studied a series of technologically ambitious ideas that were embodied in a comprehensive Air Force reconnaissance satellite effort known as WS (Weapons System) 117L. This program was formally initiated in late 1956.

During 1957, several individuals at RAND developed the concept of a technologically simpler photoreconnaissance satellite that would return exposed film to Earth in a reentry capsule rather than send it back electronically. This concept was made part of the overall WS-117L effort in the fall of 1957. This was still an Air Force program, conceived by Air Force contractors to serve what was fundamentally an Air Force mission. The WS-117L program was large and ambitious, and the film-return satellite was simply one small part of it.

In the intense re-examination of U.S. security needs that followed the Soviet Union's launch of an intercontinental ballistic missile (ICBM) in August 1957 and, much more visibly, the first artificial satellite, Sputnik I, on October 4 of the same year, the Eisenhower administration decided to move as quickly as possible to create a U.S. photoreconnaissance satellite capability. Eisenhower's advisers seized on the RAND concept for a film-return satellite, and suggested that it be pursued on a high-priority basis, separately from the WS-117L effort, in the highest possible secrecy and under different management arrangements. The intent was to have this "interim" program obtain information about Soviet strategic capabilities as quickly as possible, with the more capable satellites that were part of WS-117L replacing the interim program as soon as possible.

President Dwight D. Eisenhower approved this plan in February 1958. He decided that the Central Intelligence Agency (CIA) would have the lead role in the program, but that the program would be managed jointly by the CIA and the Air Force. This was similar to the management arrangement that had been put in place for the highly classified U-2 reconnaissance airplane program in 1954, which had worked quite well in the succeeding three years. CORONA was intended to copy that success; it was not clear at the time to anyone involved that it was anything more than a temporary program to bridge the gap until the more sophisticated Air Force reconnaissance satellites were developed.

What in 1958 was intended as an "interim" program, however, lasted 14 years. The program was given the code name CORONA; one of the individuals drafting the initial program directive chose the name as he looked down at his typewriter. (One of the leading typewriter brands was called Smith-Corona.)[4] In order to disguise the actual nature of the program, early flights were identified as part of an engineering test and biomedical science satellite program called Discoverer. In all, there were 145 CORONA launches, the first in February 1959 (only 12 months after the program had been approved) and the last in May 1972. Illustrating just how important CORONA was to U.S. security requirements and thus how determined the Eisenhower administration was to see it succeed, the program was supported through twelve mission failures before the first successful recovery of a satellite reentry vehicle in August 1960. While today many talk of the amazing success of CORONA, it is easy to forget that in the early days it was plagued by one problem after another. Thus, while some have called CORONA a triumph of American technology, the achievement came only through perseverance.

The program collectively known as CORONA was comprised of six similar but distinct satellite models and three different intelligence objectives. In 1962, a system of designating these separate models was put into use. The satellites were called KEYHOLE (indicating their use to obtain overhead imagery), abbreviated KH. Each satellite model was given a number; thus the first three CORONA models were retroactively designated KH-1, KH-2, and KH-3. At the time the new identification system was put into use, the model being developed was KH-4. Improved versions of this model were later identified as KH-4A and KH-4B. All of these satellites were intended to return film containing images of areas of strategic interest to U.S. policymakers and security planners.

Two similar satellite models were developed by the same government-contractor team and considered part of the overall CORONA program: the Army's KH-5 satellite carrying the ARGON mapping camera, and the KH-6 LANYARD wide-area "spotting" system. After twelve launches under the KH-5 designation, ARGON's objectives were subsumed into overall CORONA missions. LANYARD was basically an attempt to make use of the hardware

A CORONA film-return bucket descending after reentry, as seen from a recovery aircraft.

remaining from a canceled Air Force program (called SAMOS) that had been part of the WS-117L effort. The KH-6 was a temporary program and its early missions proved problematic; the program was halted after only three launches.

Although history often focuses on the more dramatic "firsts" of the space race, it is also important to identify the ones that represented true pioneering work in space technology. CORONA achieved a number of notable firsts: first photoreconnaissance satellite; first recovery of an object from space (and first mid-air recovery of an object from space); first mapping of Earth from space; first stereo-optical data from space; and first program to fly more than 100 missions in space. In all, some 120 of the 145 CORONA missions were complete or partial successes. CORONA resulted in the exposure of over 2.1 million feet (almost 400 miles) of film and took over 800,000 photographs.[5] CORONA photographed a total land area of 557 million square miles.[6]

Beyond the sheer physical achievements of the program were its achievements in the intelligence realm. CORONA resolved the "missile gap" question (determining that there indeed was one, but in favor of the United States). It also located *all* Soviet ICBM sites, *all* intermediate-range ballistic missile (IRBM) sites, *all* antiballistic missile (ABM) sites, and *all* warship bases, submarine bases, and previously unknown military and industrial complexes. It also provided information about the size and deployment of Soviet conventional military bases and dramatically improved mapping data on all of these targets.

The value of this information to U.S. national security cannot be overemphasized. Prior to the development of CORONA, the United States was planning to defend itself, and to fight if attacked, without knowing even the basic capabilities of its adversaries; essentially it was blind. In an era of strategic bombers and hydrogen weapons, this was a most dangerous position. Within this environment, the military leadership demanded that the country prepare for the worst-case scenario. A number of expensive arms programs were begun because of this lack of data.

CORONA removed this uncertainty. Its very first successful mission, in August 1960, immediately demonstrated the value of satellite reconnaissance. This mission returned more imagery of the Soviet Union than the twenty-four previous U-2 missions over the Soviet Union combined. It revealed sixty-four new Soviet airfields and twenty-six new surface-to-air missile (SAM) sites—all in one mission. Estimates of Soviet capabilities in the late 1950s were stated in vague terms with low confidence. Estimates of the locations of key targets were accurate to only several miles at best, and up to two dozen miles at worst. American strategic forces, particularly ballistic missiles, could not have been

A C-130 Hercules aircraft catching a descending CORONA film-return bucket. A nylon cable strung between two metal poles trailing the plane would catch the parachute lines. Crewmen would then winch the payload into the airplane. The entire recovery procedure required a highly skilled crew.

targeted accurately without overhead reconnaissance information and thus could have been rendered militarily useless.

By the mid- to late 1960s, all of this uncertainty was gone. Estimates of the numbers of key weapons were stated in precise numbers with high confidence. The geodetic coordinates of targets were known to within several hundred feet. The U.S. strategic deterrent was dramatically enhanced.

OPENING UP INTELLIGENCE FILES

As mentioned earlier, a volume of this character would not have been possible just a short time ago. On February 22, 1995, President Bill Clinton signed Executive Order 12951 on "Release of Imagery Acquired by Space-Based National Intelligence Reconnaissance Systems." That executive order, in turn, was part of a general movement toward openness in previously highly classified areas. During the last few years, the Intelligence Community has declassified and released for public examination and use vast quantities of once highly sensitive records and imagery. More material has been released to the National Archives since the mid-1990s than in all of the previous decades since the end of World War II and the beginnings of the modern U.S. intelligence establishment. Although scholars and other researchers continue to hope for even broader access to the enormous quantities of still sensitive and protected records, and some will never be satisfied with the results, many of them recognize that no other foreign intelligence service in the world has carried out a sustained and systematic declassification program like that of the U.S. Intelligence Community.

As of mid-1997, the CIA alone had released approximately a quarter-million pages of records relating to the assassination of President John F. Kennedy, about 14,000 pages of intelligence records for inclusion in the Department of State's *Foreign Relations of the United States* series, and thousands of pages of declassified articles from the professional journal *Studies in Intelligence.* Valuable intelligence histories, virtually the entire archives of the pre-CIA Office of Strategic Services, and more than 450 National Intelligence Estimates related to the Soviet Union and international communism are available at the National Archives. The National Security Agency has released all of the VENONA transcripts, a total of 2,900 Soviet intelligence messages.[7] So the declassification of CORONA and its results are just part of a larger movement.

In addition, a large volume of additional intelligence records is currently in the queue for declassification review. This material includes all of the analyses on the Soviet Union completed by the CIA's Intelligence Directorate from the late 1940s to the demise of the USSR. Records of early Cold War covert actions, including the 1954 actions against the Arbenz government in Guatemala and the 1961 Bay of Pigs invasion, have had a high priority for review and release. Future declassifications will also include the reconnaissance satellite programs that followed CORONA.

In recent years, the stimulus for declassifying historical intelligence records has generally come from one of four sources: (1) legislative mandate; (2) executive order; (3) the recommendations of an independent panel of scholars who advise the Director of Central Intelligence (DCI); or (4) wholly voluntary initiatives by the Intelligence Community. (In addition, of course, large quantities of records, both historical and contemporary, are released in compliance with specific requests under the provisions of the Freedom of Information Act legislation.)

The release of voluminous records relating to the assassination of President John F. Kennedy is a prominent example of a legislatively mandated declassification program. The Assassination Records Review Board was established by legislation that passed the Congress unanimously in 1992 and was signed into law by President George Bush. Members of the board, who have unprecedented authority to contest Intelligence Community decisions regarding the withholding of information relating to the assassination, were subsequently appointed by President Clinton.

Legislation passed in October 1991 also set out rather elaborate procedures for declassifying historically valuable intelligence records. The Foreign Relations Authorization Acts for fiscal years 1992 and 1993 specify how departments, agencies, and other entities of the U.S. government shall cooperate with the Office of the Historian at the State Department "by providing full and complete access to records pertinent to U.S. foreign policy decisions and actions and by

CORONA took its first image, pictured here, on August 18, 1960. The white strip at the center of the photo is the Mys Schmidta airfield in the northern Soviet Union. The quality of the resolution that photo interpreters and analysts had to work with originally was quite limited—resolution was 35–40 feet. Later CORONA images had resolution as good as six feet. (Photo courtesy National Photographic Interpretation Center)

providing copies of selected records" for inclusion in the widely used volumes of the Foreign Relations of the United States.

Within the Intelligence Community, the CIA has been the agency most affected by this legislation. In response, a number of steps have been taken; support for the Foreign Relations series has been considerably expanded. For example, State Department historians have been provided full access to CIA records relevant to planned volumes in the series. They also have access to CIA finding aids and, in contrast to past practice, now are using citations to CIA archives in the volumes. CIA staff historians in the agency's Center for the Study of Intelligence assist at every stage in the research and editing of Foreign Relations volumes that include intelligence records.

President Clinton's 1995 Executive Order 12958 prescribes ambitious declassification review programs and objectives that have been adopted by intelligence agencies. The CIA Center for the Study of Intelligence has responsibility for historical or systematic declassification of records that are at least 25 years old; a separate staff has been established to conduct automatic declassification reviews. The two components work closely together in order to facilitate declassification and release of the largest possible quantities of records.

A third, and increasingly important, means of identifying and recommending historical intelligence records for declassification review is the external Historical Review Panel that advises the DCI. This panel of distinguished American scholars was enlarged and began to meet regularly in 1995. During each of the panel's biannual meetings that year, members met with then–CIA Director John Deutch, and two of their initial recommendations for priority declassification review topics—DCI records from the late 1940s through 1962, and intelligence analysis on the Soviet Union from the late 1940s through August 1991—are currently being implemented.

Some of the largest and most significant declassification review priorities of U.S. intelligence agencies fall into the fourth category, voluntary review. These priorities have been devised and implemented because of the commitment of recent DCIs to greater openness by the Intelligence Community. Beginning with Robert M. Gates, who in February 1992 publicly articulated a comprehensive openness agenda, DCIs and other senior CIA officials have directed wholly voluntary releases of records and imagery. These materials are freely accessible at the National Archives and elsewhere to any interested scholar or researcher.

HOW THE CORONA IMAGERY WAS DECLASSIFIED

By far the largest single voluntary declassification project undertaken thus far has been the release of more than 2 million linear feet (800,000 photographs) of original imagery taken from space by the CORONA reconnaissance satellites. This was far from a straightforward undertaking. Any declassification review of intelligence records by definition is arduous and fraught with problems that arise from the statutory responsibility of CIA directors to protect intelligence sources and methods. For example, the identities of individuals from foreign countries who have cooperated with U.S. intelligence organizations will not be revealed, even decades after that cooperation ended. Similarly, intelligence methodologies and technical capabilities must be protected, often for extended periods of time, because compromising them could provide enemies of the United States with the ability to take offensive or defensive actions detrimental to U.S. interests or citizens.

An imposing burden is therefore placed on officials and review officers working in the various declassification programs of the Intelligence Community. On the one hand, they strive to release the largest possible quantities of historically valuable records in the prevailing spirit of greater openness. But typically they must review records word by word in order to be certain that sensitive "sources and methods" information is not inadvertently released. They must expunge names of still-sensitive sources or technical capabilities. And they must take care to be certain that seemingly innocuous pieces of separate identifying information cannot be reassembled after release of records, if such a reconstruction would be damaging to intelligence sources and methods. The results of such efforts by declassification review officers are evident in blacked-out (formally known as "redacted") portions of intelligence records available at the National Archives. In the future, of course, re-reviews of redactions may permit the release of excised words or passages.

The release of the CORONA imagery, which was completed in 1996, posed declassification challenges that in some respects even exceeded those described above. Soon after he was sworn in as DCI in November 1991, Robert Gates formed two task forces, one to consider the feasibility of declassifying reconnaissance satellite imagery, and another to determine how the Intelligence Community might use its technology to assist scientists studying the global environment. To a considerable extent, Gates's initiative was motivated by interest expressed by then-Senator Albert Gore in the possible declassification and use for environmentally related purposes of imagery taken by obsolete broad-area search satellite systems.

This was an unprecedented proposition. Since the origins of the CORONA program in the 1950s, the existence of satellite intelligence systems had remained among the most highly classified of government activities. Beginning with the Eisenhower administration, and continuing through that of Gerald Ford, intelligence analyses derived from satellite imagery of the Soviet Union, China, and other areas of the world continually informed critical policy decisions. But it was not until 1978, when President Jimmy Carter announced that the United States used satellites to verify arms control treaties, that such reconnaissance activities were publicly acknowledged.[8] And, as indicated elsewhere in this volume, it was not until September 1992 that the existence of the National Reconnaissance Office was officially revealed.

Thus, Gates's task forces faced formidable obstacles. For example, how should the United States acknowledge the fact that it took photographs of countries, both unfriendly and friendly? There was also concern that, since satellites had also on occasion taken photographs of the United States, the specter of domestic spying might be raised. Declassifying satellite imagery would necessarily mean that details of optical and satellite technology and capabilities would for the first time be released to the public. Might such releases compromise other intelligence capabilities?[9]

The release of imagery would also require the concurrence of numerous entities within the Intelligence Community; this took time. Intelligence agencies would also have to change numerous regulations, security control measures, and international handling guidelines to reflect declassification of the CORONA, ARGON, and LANYARD systems; there were also concerns about the implications of releasing the imagery for future Freedom of Information requests.[10] The declassification challenge escalated further, moreover, when Intelligence Community attorneys determined by late 1993 that only the president of the United States could order the declassification of satellite imagery.

As a result, President Clinton signed Executive Order 12951 on February 22, 1995, authorizing release of "scientifically or environmentally useful imagery acquired by space-based national intelligence reconnaissance systems, consistent with national security." The executive order authorized the declassification and release to the National Archives of imagery from the CORONA, ARGON, and LANYARD systems, and specified that the transfer be completed within eighteen months. The executive order also stipulated:

In consultation with the Secretaries of State and Defense, the Director of Central Intelligence shall establish a comprehensive program for the periodic review of imagery from systems other than the Corona, Argon, Lanyard missions, with the objective of making available to the public as much imagery as possible consistent with the interests of national defense and foreign policy. For imagery from obsolete broad-area film return systems other than Corona, Argon, and Lanyard missions, this review shall be completed within 5 years of this order. Review of imagery from any other system that the Director of Central Intelligence deems to be obsolete shall be accomplished according to a timetable established by the Director of Central Intelligence. The Director of Central Intelligence shall report annually to the President on the implementation of this order.[11]

Vice-President Gore visited CIA headquarters on February 24, 1995, for a ceremony at which the executive order was announced. He said, "Satellite coverage gave us the confidence to pursue arms control agreements—agreements that eventually led to dramatic decreases in the number of nuclear weapons and their delivery systems." He also noted that satellites "recorded much more than the landscape of the Cold War. In the process of acquiring this priceless data, we recorded for future generations the environmental history of the Earth at least a decade before any country on this Earth launched any Earth resource satellites."

Three months later, on May 24–25, 1995, a major conference cosponsored by the CIA's Center for the Study of Intelligence and the Space Policy Institute of George Washington University's Elliott School of International Affairs showcased the first declassified CORONA imagery and brought together most of the living members of the "CORONA Pioneers," those individuals who had been closely involved with the program in its early years. A volume of newly declassified documents, compiled by Kevin Ruffner of the History Staff of the Center for the Study of Intelligence, was published to coincide with the conference. It included a newly declassified article by Kenneth E. Greer that was the first history of CORONA.[12] The article was first published in 1973 in a classified issue of *Studies in Intelligence,* the journal of the foreign intelligence profession which was founded in 1955. The conference volume also included a declassified Special National Intelligence Estimate issued in September 1961, entitled "Strength and Deployment of Soviet Long-Range Ballistic Missile Forces," which contained important strategic military analysis based on CORONA imagery.

OVERVIEW

This volume goes well beyond the discussions at the May 1995 "Piercing the Curtain: CORONA and Revolution in Intelligence" conference, but it also draws heavily on discussions during those two days. In particular, chapter 2 by Albert "Bud" Wheelon, who was one of the leaders in establishing a continuing U.S. photoreconnaissance satellite program, is a revision of his conference keynote speech, and chapters 8–11 are edited versions of the discussions by conference speakers of various aspects of the CORONA program and of the use of space-derived photo intelligence. Although these portions of the volume are not proceedings in the conventional sense, they are intended to preserve the most historically valuable elements of the conference discussions.

Chapters 1 and 3–6 are original essays by individuals able to put early U.S. photoreconnaissance satellite efforts in their historical and political context. Chapter 1, by Harvard Professor of History Ernest May, details the specific

A CORONA photograph of the Soviet N-1 lunar rocket–launch complex. The N-1 competed with the American Saturn V program in the race to the moon. The fuel and oxygen tanks for the launch pads are located in the center of the photo, between the two launch pads.

impact of CORONA intelligence on strategic assessments of the Soviet Union. Chapter 2, by Albert Wheelon, discusses the general technological background and development of the CORONA satellite. Political scientist/historian Dwayne A. Day addresses the development and improvement of the CORONA satellite in chapter 3. In chapter 4 historian R. Cargill Hall elaborates fully on the origins of both the concept of strategic intelligence in the post–World War II period and the various U.S. discussions and initiatives in strategic intelligence that preceded, and ultimately led to, CORONA.

One important difference between gathering intelligence from an orbiting satellite and from an overflying airplane or balloon was that the latter was clearly a violation of territorial sovereignty and thus in almost all cases illegal under international law. Before the first satellite was orbited, there was no international agreement on whether overflying a particular nation while in orbit would be viewed as infringing national sovereignty. Chapter 5, by Dwayne A. Day, portrays the focused U.S. efforts during the mid-1950s to use a proposed civilian scientific satellite program conducted as part of the cooperative International Geophysical Year to establish the precedent that satellite overflight was *not* a

A June 1970 CORONA image of the Soviet N-1 lunar rocket–launch complex with an inset closeup of the west launch pad. An N-1 vehicle is on the west launch pad at the right, and the east pad still shows evidence of an on-pad explosion in July 1969. The N-1 was the size of the American Saturn V rocket. This explosion of the unmanned rocket occurred shortly before Apollo 11 reached the moon. While the U.S. Intelligence Community knew that the Soviets were still engaged in a lunar program as late as 1969, such information was not common knowledge and the Soviets denied that they were racing the Americans to the moon. This explosion, which devastated one of the two launch pads, was a serious setback for the Soviet lunar program. Note the scarring on the left launch pad. The baffle plates over the flame trenches and a large lightning tower have been blown away. The launch tower has also been damaged.

The inset shows the "6L" N-1 vehicle as a fat, bullet shape at the center. The booster has no L-3 payload and is capped by a temporary cone. (Inset photo courtesy of C. P. Vick.)

A CORONA stereo image of the same target—the terminal area and Theme Restaurant at Los Angeles International Airport. Stereo images provided photo interpreters and analysts with extremely important information; they gave them a sense of depth perception so that they could assess objects more accurately. CORONA first started taking stereo images with the introduction of the KH-4 system in January 1962. These two images were taken several seconds apart by the KH-4's two main cameras. Note the movement of the planes on the tarmac and the different angle of the restaurant at the center of the photos.

violation of sovereignty. This precedent would clear the legal path for the use of satellites to collect intelligence-related information.

The CORONA program was initially managed under an ad hoc arrangement involving both the CIA and the U.S. Air Force. This arrangement was patterned on a similar management structure for the U-2 program. By 1961, it was clear that the United States was likely to carry out a continuing program of satellite reconnaissance, and a more formal management structure was created. The result was the National Reconnaissance Office (NRO), an organization whose very existence was classified until 1992. In one of the first discussions of the NRO, the organization's historian Gerald Haines discusses the NRO's origins and early activities, including various institutional and personal controversies during NRO's start-up years.

Chapter 7 departs from the subject of CORONA to discuss its Soviet counterpart, known as Zenit. This chapter, written by Russian space historian Peter Gorin, demonstrates that some aspects of the Soviet space reconnaissance program were very similar to the American CORONA, whereas others were very different.

The second part of the book is based largely on the words of those who themselves built and used CORONA. As mentioned above, chapters 8–11 contain more personal views of initial U.S. photoreconnaissance efforts and specifically of the CORONA program from individuals who were intimately involved in the 1958–72 period when the program took shape and came to fruition. Here they discuss how CORONA was used by the U.S. presidents, how the system originated and evolved, how it revolutionized mapmaking, and how they utilized the CORONA photographs. The excerpts from the May 1995 CORONA conference selected for inclusion here represent only a small portion of the discussions then; each participant reviewed the elements of his or her conference presentation selected for inclusion by the editors, and many took the opportunity to revise or expand on their original remarks.

The objective of this volume is to present a comprehensive, authoritative portrait of the origins of the U.S. photoreconnaissance satellite program, and particularly of the component of that program called CORONA. Such a portrait is particularly valuable because of CORONA's significance to U.S. national security during the height of the Cold War. Perhaps Wheelon best summarizes the overall impact of the program:

> CORONA made an extraordinary contribution to world stability. . . . It guided U.S. national security policy during the worst years of the Cold War and eventually enabled arms control treaties to be negotiated and monitored with confidence—treaties that now are reducing nuclear and conventional arsenals dramatically. This first satellite reconnaissance system was truly a triumph of American technology.[13]

Part One

THE CORONA STORY

The essays in this part of the volume discuss the CORONA program from various perspectives, ranging from its place in the history of the Cold War and in the emerging importance of strategic intelligence to national security policy, to its contributions to advancing space technology. The CORONA program was, of course, not an isolated development; it was the product of the convergence of new technological possibilities with Cold War strategic needs. Both of these aspects of the program's history are fully covered in the essays included in this part of the book. The need to manage CORONA and its successor reconnaissance satellite programs in utmost secrecy gave rise to the need for organizational innovation; that story is told for one of the first times here. The value of strategic intelligence obtained from satellites was obvious to the Soviet Union as well as to the United States. Thus it comes as no surprise that the USSR paralleled CORONA's development with a reconnaissance program of its own, Zenit. An overview of that program adds a useful dimension to the book's contents.

Six of the seven essays included in part one were prepared especially for this book, though all draw on their authors' previous work. The essay by Wheelon is an adaptation of his keynote address at the 1995 conference that began the collaboration among the editors. Because many of the essays discuss the same events from their author's particular perspective, there is some duplication of information; the editors chose not to try to eliminate all of this duplication, since each essay author tells the CORONA story in a unique way. The sum total of the material in this part of the book and in part two, we believe, provides a comprehensive look at CORONA, truly an "eye in the sky" serving the U.S. national interest.

ERNEST R. MAY

1
STRATEGIC INTELLIGENCE AND U.S. SECURITY
The Contributions of CORONA

The primary significance of the first U.S. photoreconnaissance satellite effort, the CORONA program, can best be understood in the context of the U.S-Soviet strategic nuclear balance. A strong case can be made that CORONA was a major factor in shaping the character of that balance during the 1960s, and even beyond.

ASSESSING THE STRATEGIC BALANCE

In the mid-1950s, President Dwight D. Eisenhower and members of his administration began to worry about the threat posed by Soviet long-rang missile programs. Given the "New Look" strategy just adopted, the administration was trying to estimate the minimum amount of conventional military force needed to deter the Soviets from launching a ground attack against Western Europe. The "New Look" strategy presumed that, because of the deterrent effect of America's strategic nuclear capabilities, these conventional forces could be almost token in number. (In a public exposé of the strategy, Secretary of State John Foster Dulles had warned that America's response to Soviet aggression would be "massive retaliation.") But RAND Corporation studies, briefed to Washington policymakers by Dr. Albert Wohlstetter, had begun to make clear the potential vulnerability of America's strategic nuclear bomber force. A Soviet surprise attack on those forces, RAND analysts estimated, could reduce to almost zero the United States's ability to carry out the threatened retaliation. Particularly alarming was the possibility that the Soviets would stage such an attack with guided missiles as well as bombers, for missiles were much harder to defend against. Indeed, there was doubt as to whether *any* defense against missiles was feasible.[1]

The U.S. Intelligence Community responded to administration questions about Soviet missile programs with a 1954 National Intelligence Estimate (NIE). Vetted by representatives of all relevant intelligence agencies and endorsed by the Director of Central Intelligence (DCI), such an estimate was the ultimate finished product of the community. NIE 11-6-54, issued in October 1954, summarized all that the community knew about the subject as of that date. Because of its bearing on judgments about the contributions of CORONA, the foreword of this NIE deserves to be quoted at length.

> In preparing this estimate we have had available conclusive evidence of a great postwar Soviet interest in guided missiles and indications that the USSR has a large and active research and development program. However, we have no firm current intelligence on what particular guided missiles the USSR is presently developing or may now have in operational use. Therefore, in order to estimate specific Soviet missile capabilities we have been forced to reason from: (a) the available evidence of Soviet missile activity, including exploitation of German missile experience; (b) our own guided missile experience; and (c) estimated Soviet capabilities in related fields. In addition, we have analyzed such factors as: (a) Soviet industrial resources and economic capabilities; (b) Soviet nuclear capabilities in relation to guided missiles; (c) the estimated reliability of missile systems; (d) various logistic and training factors; and (e) Soviet capabilities in geodesy and cartography. Finally, in the absence of current evidence on specific Soviet missile projects, we have estimated Soviet intentions on the basis of probable Soviet capabilities in this and other weapons fields. Therefore our estimates of missile characteristics and of dates of missile availability must be considered as only tentative, and as representing our best assessment in the light of inadequate evidence and in a new and largely unexplored field.[2]

In 1954, accurate long-range nuclear-armed missiles existed only as theoretical constructs. But by the end of 1957 Soviet Sputniks 1 and 2 had demonstrated to the entire world both the feasibility of an intercontinental ballistic missile (ICBM) and the fact that the Soviet Union possessed the capability to develop such a weapon system. With photography from high-altitude U-2 aircraft complementing sources such as those enumerated in the 1954 NIE, the U.S. Intelligence Community was able to provide more detail on Soviet missile programs. The scale of Soviet efforts, however, remained extremely hard to interpret.

In the run-up to the 1960 presidential campaign and in the campaign itself, charges were constantly made that the Eisenhower administration was permitting development of a "missile gap," the result of which would be Soviet predominance in ICBMs. Although the president and others in the Eisenhower administration found U-2 photography reassuring and attempted to calm public opinion, they could not reveal the source of their confidence because of the highly classified nature of the U-2 missions. In addition, the Intelligence

An image of a Soviet SS-9 ICBM launch silo taken on September 8, 1967. American leaders were particularly concerned about the massive SS-9, but CORONA helped allay their fears by providing images that analysts could use to determine how many of the weapons were actually deployed. ICBM silos were easily identifiable because of their long access roads with gradual turns. Such roads were necessary to accommodate the tractor-trailer trucks used to transport the missiles. (Photo courtesy National Photographic Interpretation Center)

Community gave top officials comparatively unsteady support. An NIE of mid-1960 asserted: "Since there is insufficient direct evidence to establish the scale and pace of the present Soviet ICBM production and deployment program, we have based our estimate in part on various indirect forms of evidence and on argument and analysis deduced from more general considerations." Each member of the U.S. Intelligence Board had his own conclusion. The Air Force Assistant Chief of Staff for Intelligence predicted that the Soviets would have 150 ICBMs by mid-1961 and 700 by mid-1963. His Army and Navy counterparts predicted 50 ICBMs by the first date and 200 by the second. The DCI split the difference, predicting 150 and 400. The State Department representative predicted that the numbers would be somewhat higher, but not as high as those foreseen by the Air Force.[3]

In the autumn of 1961, with John F. Kennedy by then president, the Intelligence Community produced an NIE significantly different in both tone and content:

1. New information, providing a much firmer base for estimates on Soviet long-range ballistic missiles, has caused a sharp

downward revision in our estimate of present Soviet ICBM strength. . . .

2. We now estimate that the present Soviet ICBM strength is in the range of 10–25 launchers from which missiles can be fired against the U.S., and that this force level will not increase markedly during the months immediately ahead.[4]

The new specificity possible in intelligence estimates presented to the administration is further indicated by an NIE of mid-1962. It commented on levels of activity at Soviet missile test ranges, put at only six to ten the number of first-generation Soviet ICBMs, but provided details on two second-generation systems, designated as SS-7 and SS-8.[5] By 1964 the Intelligence Community could describe at length for President Lyndon B. Johnson not only Soviet missile tests in progress but the basing, flight characteristics, and probable payloads of yet another generation of ICBMs, designated as SS-9. Although the community still voiced uncertainty about probable future deployments, the range of uncertainty was much narrower than before 1961.[6] Based on the new information, U.S. security planners knew that they did not have to spend billions of dollars to develop the systems needed to counterbalance a threat of massive Soviet ICBM attacks on U.S. territory.

The change in these intelligence estimates was largely due to CORONA. Although the initial imagery obtained from this space-based reconnaissance system was inferior in quality to that from U-2s, the area covered was immensely wider. Indeed, the very first CORONA mission covered more of the Soviet Union than had all U-2 flights combined, and the imagery included areas of the Soviet Union not previously photographed.[7]

Technical improvements described elsewhere in this volume quickly made CORONA imagery a match for that from the U-2. Within a few years, CORONA was making it possible for the Intelligence Community to portray Soviet strategic missile programs with a degree of assurance hardly conceivable to authors of the NIEs of the 1950s. In June 1967, for example, the managers of CORONA presented the results of a five-day mission. The mission produced photography of all twenty-four known Soviet ICBM complexes. It resulted in an almost exact count of deployed Soviet ICBMs—between 893 and 898.[8]

THE IMPACT OF CORONA

It is not possible to state with any assurance the larger effects of CORONA. It is clear that President Eisenhower thought the program of highest priority. He persisted in pressing the development of CORONA despite an extraordinary number of expensive initial failures. We can only guess as to the course of history had he lost patience and given up and the advent of satellite imagery therefore been postponed. The actual course of history was much happier than it might have been. The rival Cold War alliances had arsenals of nuclear weapons adequate to destroy civilization, perhaps life, in at least the Northern Hemisphere. They did not use them. The sure knowledge of Soviet ICBM and other capabilities bestowed on American leaders by CORONA may have been a key reason.

The summer of 1961, when CORONA imagery was first being successfully recovered on a regular basis, was a period of intense Cold War tension. In 1958 Soviet leader Nikita Khrushchev had begun to threaten unilateral action to change the status of divided Berlin. He periodically delivered ultimatums, then withdrew them. Meeting new U.S. President John F. Kennedy in Vienna in June 1961, Khrushchev renewed his threats. Kennedy returned from Vienna fearful that Khrushchev intended this time to force a crisis. What happened was the sudden erection of the Berlin Wall, followed by gun-barrel to gun-barrel face-offs between Soviet and American tanks. In retrospect, we can see the Berlin Wall as a punctuation point, temporarily easing the problems posed for the Eastern bloc by massive emigration from East to West Germany. At the time, however, American leaders thought the building of the wall merely a preliminary to further moves aimed at testing the willingness of the West to fight for Berlin. Policy memoranda exchanged in Washington in the fall and winter of 1961–62 discussed seriously the possibility that tension over Berlin would culminate in exchange of nuclear weapons.[9]

The retrieval of the first good CORONA imagery, as reflected in NIE 11-8/1-61, produced a collective sigh of relief in Washington. Although Kennedy and members of his administration had already begun to doubt the soundness of election-time allegations of a wide missile gap, they retained qualms. CORONA demonstrated that, far from having scores of ICBMs, the Soviets had approximately six. With crash programs to field Atlas and Titan ICBMs, to develop a follow-on generation of the solid-fueled Minuteman, and to put ballistic missiles on submarines, the United States was far ahead. Although a missile gap existed, it ran very much in America's favor.

Beginning in the autumn of 1961, administration officials quietly put the Soviets on notice of this fact. Administration rhetoric, meanwhile, became more calm in tone. Some historians who have reviewed recently released records of the Kennedy White House believe that the president himself began in early 1962 to think in terms of something like the détente achieved by President Richard M. Nixon eight to ten years later.[10]

America's new CORONA-induced confidence did not have a calming effect on Moscow. Khrushchev made rash decisions that produced the October 1962 Cuban missile crisis. He had many reasons for making these decisions. A desire to offset American missile superiority was only one. Affected by Cuban pleas for protection against a possible American invasion, by worry about repercussions within the Communist bloc should the Chinese take him to task for not adequately protecting the Cubans, and by other concerns, Khrushchev could well have acted as he did, even if American self-confidence had not taken a visible upturn.

When the Cuban missile crisis actually occurred, CORONA steeled the nerves of the Kennedy administration. Imagery from CORONA and related intelligence enabled the president and his associates to judge not only whether missile sites in the Soviet Union were making launch preparations, but also whether other types of Soviet forces were moving into position for offensive operations. On the eleventh of those awful thirteen days, for example, Kennedy's intelligence advisers could reassure him that, while some Soviet bloc armed forces were increasing their operational readiness, there were still not significant redeployments.[11]

One can argue, of course, that the Soviets might have redeployed forces without necessarily intending hostilities, and that knowledge of these redeployments could have caused the United States to precipitate a still-avoidable war. But this did not happen. Day after day, reports based on CORONA continued to give President Kennedy confidence that he could prolong the effort to reach a negotiated settlement. CORONA thus contributed significantly to the dénouement of the most dangerous crisis in the history of humankind, when, on the last of thirteen days, Khrushchev drew back, publicly promising to withdraw his nuclear-armed ballistic missiles from Cuba, and Kennedy reciprocated by publicly promising not to invade Cuba and secretly promising to withdraw long-range missiles from Turkey.

In the decade after the Cuban missile crisis, CORONA played a significant role in shaping both U.S. defense policy and U.S. relations with the Soviet Union.[12] From the 1950s to the beginning of the 1960s, much American debate about national defense had focused on the appropriate number of bombers and missiles to deploy. The essential question had to do with how many strategic weapon delivery systems the United States needed in order to overmatch the Soviets. With CORONA imagery, the U.S. government could count the precise number of bombers and missiles in the Soviet inventory; this greatly simplified the process of sizing U.S. strategic forces. Before CORONA, the Air Force had proposed deploying 10,000 Minuteman ICBMs in order to be sure of having a lead over the Soviets. With information from CORONA in hand, Kennedy's Secretary of Defense, Robert McNamara, could judge the actual size of Soviet strategic forces to be such that 1,000 Minuteman missiles, less by a factor of 10, would provide a more than adequate counter. This order-of-magnitude difference saved billions of dollars.

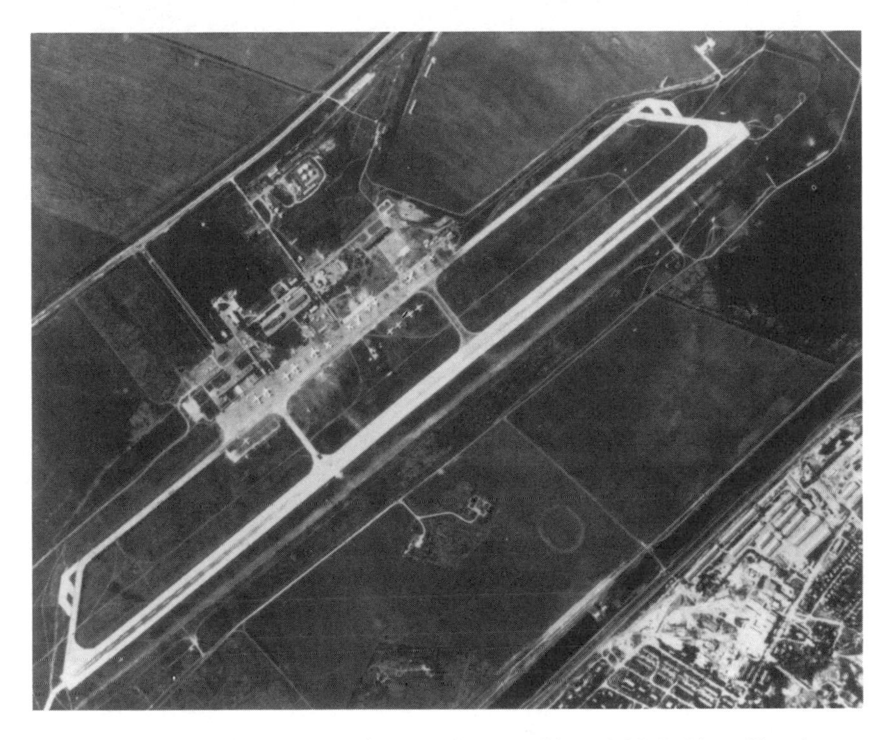

A Soviet military airfield at Mineralnyye Vody imaged by a CORONA satellite. A trained photo-interpreter could determine much information from such photographs, such as the number and type of Soviet aircraft and the presence of nuclear-weapons storage facilities.

CORONA shifted the debate about strategic forces away from quantitative and toward qualitative issues. Instead of numbers of missiles (or missile-launching platforms), planners focused increasingly on numbers of warheads and on questions concerning antimissile defense. Since data from CORONA and other sources could not easily discriminate between single-warhead and multiple-warhead missiles, planners fearful that the Soviets might gain a competitive edge turned to advocating that the United States multiply numbers of warheads and those numbers increased dramatically. By May 1972, when the United States and the Soviet Union reached their first agreement on limiting strategic arms, the United States fielded about three times as many warheads as launchers.

Since CORONA and related sources could identify and locate Soviet surface-to-air missiles but could not ascertain whether they were antiaircraft or antiballistic missiles, planners in the United States focused increasingly on proposals for U.S. antiballistic missile (ABM) systems. However, they encountered effective opposition, for it was argued that ABM deployments would threaten prospects for stabilizing the U.S.-Soviet arms competition.

Here one sees an effect of CORONA almost as important as its effect on crisis management and strategic force planning. CORONA, by creating certainty regarding numbers of deployed missile launchers, made it practicable for the United States to propose negotiated agreements limiting that category of strategic weaponry. It had become clear that the strategic nuclear arsenals of two sides were already, or soon would be, so powerful as to threaten mutual annihilation. Some argued that the two could find safety in agreements leaving each capable of destroying the other, the presumption being that knowledge of this fact would deter each from initiating war against the other. In view of the authoritarian political system of the Soviet Union and its extreme secrecy, American leaders dared not trust simply to Soviet promises. (Even if presidents had been willing to take chances, the Senate was unlikely to do so; the Senate had to ratify any arms limitation treaty.) Owing to CORONA, however, negotiations for stabilizing the numbers of launchers were possible, for the numbers could be verified.

CORONA, of course, served many other uses. It informed U.S. leaders about China's progress in nuclear weaponry. It allowed them to keep exact track of the Middle East war of 1967. It provided immense new knowledge about the earth's geography and the location of natural resources. Probably, however, the best one-line epitaph for CORONA would read: "It helped keep peace in the nuclear age."

ALBERT D. WHEELON

2
CORONA
A Triumph of American Technology

The story of the U.S. intelligence satellite program begins on December 7, 1941, when the concept of a surprise attack on U.S. territory became a vivid reality. The shock of that disaster remained forever tattooed on the memories of people who lived through it. Neither Harry Truman nor Dwight Eisenhower could forget it.

Truman believed that adequate intelligence data had been available in 1941 to warn of pending attack, but that it had been segmented and scattered. In 1945, as president, he took the initiative to establish a peacetime intelligence service. The first task of his small Central Intelligence Group was to bring all data to a sharp focus and to ensure that policy positions of the various departments of government did not color the results.

Eisenhower also remembered Pearl Harbor. During World War II, he insisted that all available intelligence be provided to his headquarters in order to plan his military campaigns in North Africa and Europe. As president, Eisenhower became increasingly concerned about the possibility of a surprise nuclear attack. The USSR had moved quickly to establish a nuclear arsenal. The threat of a swift strike became a real possibility when Soviet long-range bombers became operational.

In March 1954, President Eisenhower asked James Killian, the president of the Massachusetts Institute of Technology, and his scientific colleagues to address the problem. The resulting study was called the Technological Capabilities Panel. One of its three committees focused on the problem of strategic intelligence. Edwin "Din" Land led that work. This unique and gifted man went on to guide U.S. reconnaissance activities for the next three decades. The Killian group focused much of its attention on the need for good strategic intelligence, because it believed that such intelligence was the most highly leveraged component of national security. Din Land described the group's commitment: "We

President Harry S. Truman and General Dwight D. Eisenhower. Both men were profoundly impressed with the need for intelligence; they wanted to guard against another surprise attack by America's adversaries like Pearl Harbor. (Photo courtesy of the National Archives and Records Administration)

simply cannot afford to defend against all possible threats. We must know accurately where the threat is coming from and concentrate our resources in that direction. Only by doing so can we survive the Cold War."[1]

The Killian group enjoyed Eisenhower's complete confidence, and he ensured that they were given access to all facets of American intelligence. The group was not impressed by what they found. Their final report said:

> We must find ways to increase the number of hard facts upon which our intelligence estimates are based, to provide better strategic warning, to minimize surprise in the kind of attack, and to reduce the danger of gross overestimation or gross underestimation of the threat. To this end, we recommend adoption of a vigorous program for the extensive use of the most advanced knowledge in science and technology.

Killian and Land briefed the president personally on the specific technologies they had in mind. The U-2 spy plane was the first result of their strong influence on American presidents and American intelligence activities.

MOTIVE FOR SECRECY

Eisenhower proposed his Open Skies concept at Geneva in July 1955, only to have it rejected by Khrushchev. The Soviets protested publicly and loudly when

American reconnaissance balloons began to drift across the USSR in 1956. Vigorous diplomatic protests were made following each U-2 overflight, although the Russians never admitted to their own people that Soviet territory was being systematically violated. Eisenhower was determined not to confront the Soviets with the reality of our ongoing reconnaissance program. He refused to contradict John Kennedy's claim of a missile gap during the presidential campaign of 1960, even though he knew from both U-2 and (after August 1960) CORONA flights that the claim was wrong. He tried in every way to legitimatize overhead reconnaissance and hoped gradually to gain Soviet acceptance of it. This was a slow process, but it was finally successful. Overhead reconnaissance is now enshrined in each of our strategic AVMS limitation treaties with the former USSR, where it is described as "National Technical Means."

EARLY SATELLITE PLANNING

In 1945 the U.S. Air Force commissioned RAND to examine the feasibility of launching satellites for military purposes. RAND's early studies explored the idea of reconnaissance from space and identified the polar orbit necessary. RAND technical people, working with a growing group of Air Force officers, advocated satellite reconnaissance and did a great deal to establish the concept as a reasonable one. American ballistic missiles then being designed offered the means for lifting such satellites into orbit. RAND's work was focused primarily on the requirements of the Strategic Air Command (SAC), which was then the dominant voice in the Air Force. The SAC need was the information required for poststrike bomb damage assessment. This would allow targets that were missed by the first wave of SAC bombers to be retargeted for subsequent attacks. The resolution required for this task was not great, since one would be trying to locate large craters in relation to cities and military bases. On the other hand, this information would be needed promptly. These requirements were addressed in the satellite system proposed by RAND in 1954.[2] Its solution was based on a television satellite with a resolution of approximately 100 feet, operating 300 miles above the earth. These proposals became the blueprint for the photographic reconnaissance component of what came to be called the WS-117L program.

REFOCUSING THE SATELLITE PROGRAM

The accelerated development of American and Soviet ICBMs gave special urgency to Eisenhower's fear of a surprise attack. With only a 30-minute flight time and no realistic prospect for defense, the emerging reality of a Soviet ICBM force put the possibility of surprise attack at the top of the president's concerns. The shock of Sputnik crystallized both public and presidential concern in October 1957. It also made people in high places think seriously about satellite reconnaissance.

In 1956, the CIA's U-2 program had been expected to operate only for a year or two. In 1957, its second year of overflights, missions were few in number and limited in coverage. They could not survey large areas of the USSR to look for deployed ICBMs, and it was unrealistic to rely on them for guiding national security policy in the nuclear missile era.

There was a sense of extraordinary urgency in getting good pictures of the entire USSR; a satellite reconnaissance system was the obvious way to do so. Eisenhower asked the President's Board of Consultants on Foreign Intelligence Activities (PBCFIA) to review the Air Force program and make recommendations. The PBCFIA's report in late 1957 was skeptical that WS-117L could provide the needed capability. Because its resolution was more than 100 feet, it would be too crude to provide strategic intelligence. The satellite's resolution was limited by the small focal length of its camera and the narrow bandwidth of its television downlink. The program was running late and encountering technical difficulties.[3] In an effort to develop budgetary support for it, the Air Force had started to publicize the WS-117L program.

The U-2 prototype at its desert airstrip. This was an enormously successful project and established the precedent of close CIA–Air Force cooperation on development of reconnaissance programs. (Photo courtesy of Chris Pocock)

Killian and Land were frustrated by what they saw of the Air Force program. They strongly recommended simplifying and accelerating satellite reconnaissance activity. They wanted to start a new program based on film return from orbit, focusing on peacetime national intelligence objectives rather than on reconnaissance after a nuclear exchange. Eisenhower agreed. He was concerned with preventing nuclear war, not waging it.

Killian and Land wanted to streamline both the program and its technical management system. They urged the president to assign the leadership for a new system to the CIA, to be supported by selected elements of the Air Force. Richard Bissell would lead a small group of CIA and Air Force people, thus emulating the successful partnership that had created the U-2. This recommendation was made for several reasons. The CIA had demonstrated an ability to maintain tight security during the development phase of the U-2 project. Bissell and his CIA people had shown a remarkable capacity for moving rapidly from concept to operation. They had demonstrated an ability to make and implement decisions quickly. In addition, they were quite open to suggestions from Land and his colleagues.

The presidential decision to proceed with what became CORONA on this basis was made just eight weeks after the PBCFIA report was submitted.[4] The goal was to achieve a resolution of 25 feet or better within one year. Elements of the WS-117L program that promised early capability were transferred to the new program. As it turned out, what was imagined to be an interim system became the backbone of U.S. intelligence collection capability for the next 12 years.[5]

A second reconnaissance program was approved by President Eisenhower at about the same time. The CIA and Bissell were authorized to develop an aircraft as a successor to the U-2 that would fly at three times the speed of sound. This piloted reconnaissance system would complement the CORONA satellite. It would provide greater coverage flexibility and greater resolution than could be obtained from Earth orbit. The program was called OXCART. It became operational in 1966, but was never used over the USSR because of the pledge to stop overflying Soviet territory with airplanes which Eisenhower made following the downing of Gary Powers's U-2 in May 1960. Follow-on versions of this remarkable reconnaissance airplane were operated by the U.S. Air Force as the SR-71 until 1990.[6]

DIFFICULT DESIGN CHOICES

Richard Bissell chose Air Force Brigadier General Osmond Ritland to be his deputy for the CORONA project. The two had worked together on the U-2 program and had the utmost confidence in one another. A program office was assembled in Los Angeles in early 1958, comprised of about five Air Force officers commanded by Colonel Lee Battle. It was supported by a small group of CIA officers in Washington reporting directly to Bissell.

The program group decided to use the Thor intermediate-range ballistic missile as the first stage of a rocket combination to place CORONA in Earth orbit. This was logical because the Thor had been flying successfully since 1957 and thus was further along in development than Atlas or Titan. Thor was then in large-scale production and was being deployed to operational sites in England. More important, a Thor launch training site being established at Point Arguello in California[7] could fire directly south, and thus place a satellite in polar orbit. Douglas Aircraft Company built the Thor rocket and became a charter member of the CORONA team.

The CORONA payload needed to be placed in what came to be called "reconnaissance orbit." This is a retrograde 90-minute orbit that passes over the North and South Poles. It has a perigee of approximately 100 miles and an apogee of 240 miles. The rotating Earth turns beneath this fixed orbital plane, presenting a new swath of territory on each pass. The Thor would burn out at an altitude of 70 miles, well short of the speed required to reach this orbit. An additional rocket stage was thus needed to lift CORONA.

The Agena upper-stage vehicle had been in development at Lockheed for two years, but had not yet flown.[8] It was five feet in diameter and used a 16,000-pound rocket engine that burned storable propellants: hydrazine (UDMH) and nitrogen tetroxide. Agena was the logical choice to provide the additional velocity needed to reach Earth orbit. Lockheed thus became the second team member.

Bissell and Ritland decided to combine the second-stage and the orbital-spacecraft functions. This meant that the camera and film-recovery systems would remain attached to the Agena even after it had exhausted its fuel. Agena would have to provide precise attitude control, battery power, and thermal protection for the reconnaissance payload for several days, and later for several weeks. The choice of a camera system would determine the performance required from this redefined Agena.

The most important decision facing Bissell and Ritland was the type of camera to be used. RAND and Lockheed had done some work on a film-recovery system based on a spinning spacecraft to provide stability and to achieve the scan. It used a relatively short (12-inch) focal-length camera design by Fairchild, which had no provision for orbital image motion compensation.

An experienced reconnaissance camera design team had recently left Boston University to form the Itek Corporation. The team proposed a 24-inch panoramic camera, like those used to take pictures of large university and high school classes. The film would be held stationary in a cylindrical platen as a rotating "telescope" (i.e., a lens assembly) scanned a slit image over it. The next section of film would then be moved forward along the platen while the telescope returned to the starting position and the process was repeated. Scanning the rotating telescope over a 70-degree arc would map out a strip 10 miles by 120 miles on the ground. The specific design was based on the cameras Itek had built previously for covert balloon flights.[9] This design offered significant future opportunities for improving resolution. It was entirely mechanical, but included fixed-image motion compensation set for the planned orbital speed and altitude. The telescope lenses were diffraction limited and required precision optical glass. Bissell chose this design with the overriding objective of obtaining high-definition peacetime intelligence. With this choice, the 25-foot resolution goal was easily met; it would eventually improve to six feet.

The camera decision shifted a large burden to the Agena spacecraft. It did so at a time when the United States had virtually no experience in building such a vehicle. The panoramic camera design required that the spacecraft be stabilized around all three axes with an active control system.[10] An attitude control system using gyros, infrared horizon scanners, and cold gas jets eventually gave an accuracy of 0.20 degrees in pitch, roll, and yaw. This active control was augmented by horizon, star, and framing cameras that recorded the instantaneous vehicle attitude for later use by photo-interpreters. This decision required redesign of the spacecraft and accounted for many early development problems.

Three decades of space missions have used the basic Agena developed for CORONA. The LANYARD spotting system and ARGON mapping missions could not have been completed without it. In a very real sense, CORONA pioneered all subsequent satellite reconnaissance programs.

A time-lapse exposure of a HYAC balloon reconnaissance camera as the swing arm travels along the curved film platen, exposing the film. The 12-inch HYAC camera served as the model for the 24-inch CORONA camera. (Photo courtesy of the National Reconnaissance Office)

A fine-grain film was needed to fulfill the resolution promised by the Itek camera. Eastman Kodak had developed such film for the U-2 program. For CORONA, Eastman Kodak developed an acetate-based 70-millimeter film which was approximately three millimeters thick. It was relatively slow but gave 280 line pairs per millimeter over the entire field of view, at high contrast.[11] This compares with the best reconnaissance film used in World War II, which gave 10 to 50 line pairs per millimeter.

There was concern that scintillation induced by the turbulent atmosphere might limit CORONA resolution. Image motion of one or two arc seconds was consistently measured by astronomical telescopes.[12] At a slant range of several hundred miles, this effect could produce image smearing of three or four feet. That was below the CORONA threshold, but scintillation was considered carefully for follow-on satellite systems.[13] Fortunately, image quivering is considerably reduced when the observer is well removed from the turbulent region rather than being immersed in it, as terrestrial telescopes are.

When the CORONA program began, American ballistic missile programs had already developed ablating nose cones that could withstand the enormous heat loads generated during ballistic reentry. A capsule had never been returned from orbit, but it was clear that the heating involved in orbital reentry would be substantially less severe than a missile warhead because the satellite return trajectory is quite shallow compared to a ballistic path. The General Electric Company had developed ablation technology for U.S. ICBMs and was assigned the task of developing the film-return capsule for CORONA. The center space of the capsule held a large take-up reel for the 70-millimeter film that would pass through the Itek camera. A small rocket motor was fitted to the rear of the capsule. Agena would re-orient itself at the end of the mission and separate the capsule that now contained the exposed film.[14] The capsule would "spin up" for attitude stability, and then fire its rocket motor to reduce its speed by 1,300 feet per second.[15] This was enough to send it back to earth after traveling a quarter turn in orbit.

To begin the recovery operation, a microwave command signal was sent to the Agena as it came over the North Pole heading toward the equator. If all worked according to plan, the CORONA capsule would splash down near Hawaii in an impact area of 150 by 400 miles. The maximum heating rate during reentry would occur at 350,000 feet altitude, where the ablation heat shield would reach temperatures of 4,000 degrees Fahrenheit. A parachute would deploy when the capsule reached 50,000 feet, and would slow the descent rate to about 30 feet per second. A fleet of Air Force C-119s were deployed from Hawaii to snatch the descending capsule. These planes each towed a long nylon loop with which the air crew attempted to snare the parachute and then reel it into the aircraft.[16] Ships and helicopters were also deployed to recover the capsule if it was missed and fell into the sea. It took a great deal of trial and error to make aerial recovery a routine operation.

A two-page work statement dated April 25, 1958, reflects these daring decisions. Subsystems were rapidly built up by teams of engineers and technicians working under pressure comparable to wartime conditions. The cameras, film, and reentry capsules were integrated at a secret facility near Palo Alto, California. The completed CORONA payload was then taken to the Vandenberg Air Force Base where it was mated to the Thor-Agena rocket. The first CORONA mission was ready for launch on February 28, 1959—less than 10 months after program go-ahead.

DEVELOPMENT PROBLEMS

The first twelve CORONA missions were failures. It is important to remember that the country had almost no experience in developing satellites before 1958. The pioneering role fell on CORONA, not the much less ambitious scientific satellite program carried out for the International Geophysical Year. In the current state of technical accomplishment, we tend to forget how desperately inexperienced we all were in those early days. Vanguard had been a profound embarrassment. Thor and Atlas had gone through similar problems when they began. Titan was then experiencing repeated failures. I had worked on ballistic missile programs in Los Angeles; our monthly program review meetings were properly called "Black Saturdays." Upper stages were a new challenge. Our group had tried and failed three times to build an upper stage for Thor which could send a payload to the moon.

Getting the film capsules to land in the desired recovery areas also proved to be difficult. We found that we still had a great deal to learn about snagging the parachutes of descending film capsules. The acetate-based 70-millimeter film broke several times during orbital operations; this represented a genuine threat to the program. Fortunately, for other applications Eastman Kodak developed a polyester-based film which solved that problem.

One of the greatest problems was the pressure to continue launching at a rapid pace. This was driven by the extraordinary urgency of getting firm evidence of Soviet strategic deployments—evidence that could only come from CORONA. During the first year, despite major technical problems we launched about once a month. This short interval between launches did not give the engineers enough time to analyze and fix problems before the system was launched again. (It is useful to recall that the space shuttle was grounded for almost three years after the Challenger accident. That luxury was not available to the men and women working on CORONA.)

These problems were further compounded by the fact that the satellite had a razor-thin weight margin and thus could carry very little test instrumentation. This meant that the engineers usually did not have enough diagnostic data to correct the problems with confidence. The choice each time was simple: carry film or carry instruments. We almost always chose film. What is remarkable is that Bissell and Ritland pressed on despite these failures, and that Eisenhower continued to support them.

IMPACT ON INTELLIGENCE

The thirteenth CORONA mission was launched on August 10, 1960, and became the first completely successful flight. However, it carried no film because the program office had decided to fly a full load of diagnostic instruments. The fourteenth CORONA mission was launched a week later and, after seventeen passes over the USSR, returned 16 pounds of film.[17] This mission produced a cornucopia of data and gave more coverage of the USSR than all prior U-2 flights combined. For policymakers and intelligence analysts alike, it was as if an enormous floodlight had been turned on in a darkened warehouse.

CORONA photography quickly assumed the decisive role that the Enigma intercepts had played in World War II.[18] When the American government eventually reveals the full range of reconnaissance systems developed by this nation, the public will learn of space achievements every bit as impressive as the Apollo moon landings. One program proceeded in utmost secrecy, the other on national television. One steadied the resolve of the American public; the other steadied the resolve of American presidents.

National leaders needed a great deal of information about almost every part of the USSR. The Soviet space program had embarrassed Eisenhower with Sputnik, and would soon embarrass Kennedy with human space flights. The Soviets had a vigorous missile program and ICBM silos were reportedly being built in many places. Missile production was widely scattered. The Soviet air defense system was vast and the Soviets were developing an ABM system to counter American long-range missiles. The Soviet nuclear weapons program was vigorous and the Soviets exploded a usable hydrogen bomb before the United States.[19] An extensive chemical and biological weapons program was operating, a fact since confirmed. American intelligence could not monitor this vast enterprise with only agents on the ground. While important during the Second World War, communications intelligence was unable to identify and follow such activities. CORONA provided a unique solution to the problem.

The CIA's National Intelligence Estimates (NIEs) of Soviet strategic forces were surrounded by uncertainty and disagreement before CORONA. With very little hard data, it was possible for "hawks" to argue that the Soviet threat was enormous, while "doves" could maintain that it was trivial. This situation changed completely as satellite photography poured in. There now could be little debate about the number of Soviet bombers and missiles. Six months before the first successful CORONA mission, an NIE predicted that there would be 140–200 Soviet ICBMs deployed by 1961. One month after the first flight, that estimate dropped to 10–25.[20] Satellite photography quickly reduced the range of debate and the uncertainty surrounding these judgments. Based on CORONA images, NIEs rapidly took on the air of authority.

Perhaps just as important was the fact that anyone could understand and evaluate the photographs. Analysts and presidents alike could see large facilities and make their own judgments. The skepticism that often clouded reports from covert agents was no longer relevant. Neither was the uncertainty that surrounded the decrypting of coded messages. CORONA data was user-friendly.

The photo-interpretation community was originally sized to cope with sporadic U-2 flights. When CORONA began providing monthly coverage, this community went into overload. The CIA moved quickly to expand the cadre of photo interpreters and consolidated existing capabilities in a new National Photographic Interpretation Center (NPIC) in Washington, D.C. The CIA began developing automated pattern recognition machines that could help the photo-interpreters do their job. (It is a great satisfaction to me that these machines are now being used by the medical community to interpret mammograms.)

MANAGEMENT PROBLEMS

The streamlined management approach that Bissell and Ritland adopted was essentially the same system they had used on the U-2. It was characterized at first by seamless cooperation among the CIA, the Air Force, and the industrial contractors: Lockheed, Itek, GE, Kodak, and Douglas Aircraft. Each team member did what it did best. The extraordinary security in which the program was wrapped meant that there was no likelihood that personal or corporate recognition could be expected. Indeed, it was not until 1995—35 years later—that the CORONA Pioneers received the thanks of a grateful nation.

CORONA also had a profound influence on the future direction of the national reconnaissance program. There was a period of management tension during the early 1960s.[21] Those problems were celebrated in at least one popular book, although that account is quite confusing.[22] It is probably important for the historical record to clarify the matter from the perspective of one who was centrally involved.

A Douglas-built Long Tank Thrust Augmented Thor (LTTAT) rocket used to launch KH-4A and KH-4B CORONA satellites. The three solid rocket motors added around the base significantly increased the power of this version of the Thor. (Photo courtesy of the U.S. Air Force)

It is significant that the National Reconnaissance Office (NRO) did not exist during the development and early operational phase of the CORONA program. Nor did it exist during the first five years of the U-2 program. It had not been created when the Mach 3 OXCART aircraft started development. In each case, the primary responsibility was given to the CIA, with strong support from the U.S. Air Force. This successful partnership caused Land and Killian to push for a permanent collaboration between the two organizations. They wanted to institutionalize the extraordinary partnership that had worked so well. The NRO was thus born. (This is explained in more detail in chapter 6.)

Richard Bissell was the acknowledged leader of reconnaissance activity as the NRO concept began to take shape. However, he had taken on other important new responsibilities at the CIA. In the early 1960s, Bissell was responsible for both the reconnaissance program and the clandestine services. In the second role, he was centrally involved in preparations for the invasion of Cuba by an exile force that President Dwight Eisenhower had authorized. With the full authority of President John Kennedy, the operation proceeded in early 1961 at the Bay of Pigs, with disastrous consequences for all parties. Richard Bissell, as well as Allen Dulles and his deputy, Air Force General Charles P. Cabell, were forced to leave government service soon thereafter.[23] Herbert Scoville assumed Bissell's responsibility for CORONA, the U-2, and OXCART at the CIA. Scoville did not—indeed, could not—enjoy the standing that Bissell had created through his remarkable accomplishments. A significant vacuum was thus created in the national reconnaissance program.

The Department of Defense (DoD) moved forcefully to fill this vacuum. Secretary of Defense Robert McNamara was working to consolidate all military space activities under the Air Force. Many people in the DoD viewed the unique relationship between the CIA and the Air Force as an anomaly. They wanted their own people to take the lead in overhead reconnaissance. Joseph Charyk was held over by the new Kennedy administration as Undersecretary of the Air Force. When Bissell left, the emerging NRO role was assigned to Charyk as a second responsibility. However, John McCone became Director of Central Intelligence in 1961, and his agreement was needed to move the center of gravity from the CIA to the DoD.

The differences between McNamara and McCone were both substantive and bureaucratic. They differed profoundly on Vietnam: McNamara predicted success, while McCone was less sanguine. Then, as now, information was the coin of the realm in Washington, and McNamara wanted to control its flow. He told McCone bluntly that the CIA should confine itself to defining requirements, perhaps doing some advanced research, and being content to examine the film that Air Force satellites produced. During the Cuban missile crisis, McNamara's deputy, Roswell Gilpatric, insisted that the long-running CIA U-2 missions be transferred to Air Force control. The responsibility for Cuban overflights reverted to the CIA soon after the crisis was over, but the first Pentagon challenge had been successfully mounted.

McCone and McNamara seemed unwilling to engage one another directly on the issue of overhead reconnaissance. McCone was personally close to Gilpatric. Gilpatric took on the job of convincing McCone that the leadership of reconnaissance should pass from CIA to the DoD. McCone agreed to transfer financial control of all CIA reconnaissance activities to the NRO. Charyk used this and other concessions to shift the CIA's responsibilities progressively to his own staff.

Brockway MacMillan followed Charyk as director of the NRO and continued the policy of reducing the CIA role. In doing so, he enjoyed strong support from Eugene Fubini of the DoD staff. Their first priority was reducing the CIA's role in CORONA. This was a particularly rich plum. CORONA was highly visible and successful within the national security community, in contrast to DoD space programs. Both the SAMOS reconnaissance program and the

ADVENT military communications satellite program had been conspicuous failures, and were then in the process of being canceled by the DoD.

McCone had once served as Undersecretary of the Air Force and held its people in high regard. He was unsure whether he should support a continuing CIA role in reconnaissance. After a year of indecision and frustration, Herbert Scoville, McCone's deputy for these activities, resigned in protest.

McCone asked me to take his place and to carry on the work that Bissell had pioneered. I first came to the CIA in June 1962 to replace Scoville as director of the Office of Scientific Intelligence, as he moved over to take Bissell's role in reconnaissance. I was drawn into reconnaissance matters by McCone and Scoville because I had played a pioneering role in the Air Force ICBM and space programs. I had a high regard for both men, but noted that the objectives of DoD and CIA were no longer aligned. It seemed clear to me that the DoD saw no important role for the CIA in satellite activities. I told McCone quite bluntly: "There is no point in screwing another light bulb into a socket that is shorted out." He asked me to analyze the situation and to make a recommendation.

I concluded that the nation needed CIA participation in these vital programs. If the previous partnership was no longer possible, the country needed the benefits of competition in strategic reconnaissance. I reminded McCone that while he was chairman of the Atomic Energy Commission, the country had made a similar decision when it established a second nuclear weapon design laboratory at Livermore at the Air Force's insistence. I suggested that overhead reconnaissance was even more important than nuclear weapons design. I reminded him that the Pentagon had a sorry record in satellite reconnaissance. I argued that we could not depend on the Air Force to be the sole provider of the most important information that the country needed. I acknowledged the problems of competition among government agencies, but stated my firm belief that the overriding priority for the country was a vigorous and effective reconnaissance program.

John McCone considered this matter for several days and then decided that the CIA should again play a strong role in these programs. He pledged his full support to that objective. With his assurance and encouragement from Jerome Wiesner, Kennedy's Science Advisor, and Din Land, I agreed to pick up where Scoville had left off. I immediately assumed responsibility for all overflight activities with the U-2 and for OXCART. I also became responsible for the CIA role in CORONA.

With Scoville's resignation, McNamara's people felt that the time had arrived to move rapidly to consolidate reconnaissance responsibility in the Air Force. Brockway MacMillan, who had succeeded Charyk, tried to implement this strategy. He notified me that he was transferring CIA responsibility for CORONA payload integration and testing at the Palo Alto facility to the Air Force. Using the fiscal control that McCone had granted his predecessor, MacMillan refused to provide the funds that we normally used to reimburse Lockheed for this vital work. McCone's new policy was thus put to an early test. The CIA refused to transfer its responsibility, and the matter went unresolved for almost one year while we all debated it. During this period, Lockheed funded CORONA activity with its own corporate money rather than take sides in a bitter contest. The matter was finally resolved in the CIA's favor. Lockheed was reimbursed for the costs it incurred, and the management of the CORONA program was returned to the previous arrangement. After a period of readjustment in Pentagon expectations, the partnership between the CIA and the Air Force resumed within the NRO and served the country well to the end of the program in 1972.

The debate between CIA and DoD then shifted to whether the CIA ought to pursue new reconnaissance systems. Gene Fubini, Assistant Director of Defense Research and Engineering, and Brockway MacMillan, the Undersecretary of the Air Force and director of the NRO, vigorously opposed each new system that the CIA started to develop. The debate continued until they both left government service in 1965 and Al Flax became director of the NRO. Strongly influenced by the Land Panel, he saw the CIA and the Air Force as valuable, complementary assets. This was fortunate for the country, because the systems then being developed at the CIA would become vital components of the overall National Reconnaissance Program.

PROGRESSIVE IMPROVEMENTS

The first two years of CORONA's development were marked by great daring, repeated disappointment, and finally by extraordinary success. Having dared so much at the outset, the program settled into a pattern of gradual improvement. Performance of the Thor-Agena combination was steadily increased by extending the fuel tanks and adding strap-on solid rockets to the Thor. The larger payload capability was used to increase film load and to extend the time on-orbit from four to ten days. When the lifting capability of Thor and Agena was great enough, a second camera was added to provide stereoscopic coverage.

The camera itself was improved by going from a Tessar f/5 system to a Petzval f/3.5 lens design. The attitude control capability and vibration levels of the Agena improved with time. By these means, the resolution was gradually improved from 25 feet to 6 feet. Reference cameras were added to give the photo-interpreters very precise indications of vehicle orientation during camera operation. Color and infrared film were tried, but seemed not to increase the intelligence value when weighed against the accompanying loss in resolution and area coverage.

A KH-4A CORONA spacecraft at Lockheed's Advanced Projects facility. The KH-4A was the most prolific and successful of the CORONA designs. The two dark patches on the satellite's skin at the center of the photo are covers for the horizon sensors. This satellite was significantly larger than earlier versions because it included a second satellite recovery vehicle. (Photo courtesy of A. Roy Burks)

Recovery techniques were perfected, and it became a rare event when a film capsule was lost. The number of capsules was increased from one to two, so that one load of film could be returned and processed while the satellite continued taking pictures in orbit.

A major challenge to the program occurred in 1962. Some of the returned film was completely exposed—apparently by a bright light source. We judged that it was caused by static arcing within the satellite. Coincidentally, this electrical discharge is called "corona" by the physics community. I persuaded Professor Sidney Drell to take leave from Stanford in 1963 to lead a team of engineers and scientists in addressing this problem. In collaboration with Itek engineers, they traced the problem to the rubber rollers that were used to move film through the camera.[24] This was the first time that a younger, second-generation group of scientists became deeply involved in reconnaissance activities. Scientists like Sid Drell, Richard Garwin, Bill Perry (later Secretary of Defense), Joe Shea, and Frank Lehan went on to make enormous contributions to all U.S. reconnaissance programs during the next three decades.

NEXT STEPS

As managers in the Intelligence Community became comfortable with CORONA operations, the CIA began to look ahead. It was apparent that photo-interpreters were having difficulty locating strategic and military facilities in the vast amount of 70-millimeter film that was arriving. They felt that much finer resolution was needed if we were to increase confidence and throughput. I asked the Drell group to examine this problem in 1963, and put two questions to them: What resolution do the photo-interpreters need to find and identify strategic installations in broad-area coverage? And, could CORONA be improved to provide that level of resolution, or must we begin a new system?

The Drell team approached this problem by preparing simulated photographs from very high-quality U-2 coverage. These scenes were degraded to various resolution levels and given to the photo-interpreters to see how recognition varied with resolution. The experiments showed that a substantial improvement in resolution was indeed needed. The Drell group judged it unlikely that the CORONA system could be pushed to that new level by product improvements. CORONA's basic camera design had inherent limits—and they had been reached.

CORONA VULNERABILITY

I worried a great deal about the vulnerability of CORONA from my first days in government service. I recognized that it was an easy matter for the Soviets to identify CORONA missions and to predict their orbits. The United States was obliged by international agreement to notify the United Nations of each launch, although not of its purpose. The Agena spacecraft radiated four microwave signals on frequencies ranging from UHF band to S band. These signals carried telemetry data, but were also used by U.S. ground stations to track the spacecraft. It was clear that the Soviets could track CORONA missions using the same signals and undoubtedly knew their orbits almost as well as the United States did.

It is a relatively easy matter to destroy satellites in low earth orbit if there is an incentive to do so.[25] The United States established two antisatellite systems in 1963 and kept them in operation for almost a decade. One was based on the Nike-Zeus ABM system and was deployed on Kwajalein Island. The other used a Thor rocket on Johnston Island. The plan was to wait until a Soviet satellite passed reasonably close to these islands and then to launch with nuclear warheads to destroy it—if the command was given. The Soviets were deploying a nuclear-tipped ABM system at scores of sites around Moscow, and had several launchers at their development site near Sary Shagan. I knew that it would be a simple matter for these systems to eliminate CORONA. Had they done so, the United States would have had to conduct its affairs in almost total ignorance of Soviet activities, as we had done prior to 1960.

This image of the Pentagon, taken by a KH-4B satellite on September 25, 1967, represents some of the best CORONA imagery. The negative of this image, when viewed under a stereo microscope, would be considerably sharper than this picture. (Photo courtesy National Photographic Interpretation Center)

The destruction of a low-altitude satellite does not require nuclear weapons. Because it travels in earth orbit at speeds of approximately 17,000 miles per hour, it is only necessary to stand in the satellite's path to destroy it. American apprehension was increased in 1967, when the Soviets began to test a co-orbital anti-satellite system that could do just this.[26]

The CIA and the Air Force jointly examined a wide range of defensive measures that might protect CORONA. They considered inflating and deploying decoy balloons for the primary spacecraft. This solution had a fundamental flaw, because the balloons, following the laws of celestial mechanics, would periodically rendezvous with CORONA. Another option evaluated was orbital adjustment maneuvers that could change arrival time over defensive installations. This would require the satellite to expend fuel and to change camera sequences. To make a meaningful orbit change, one must use large amounts of fuel because the basic orbital speed is so high. Such corrections can be made only once or twice, and they would not defeat a determined attack.

None of these proposals were implemented, primarily because they required using a great deal of total payload weight. CORONA's managers opted each time for increased film loads and hoped that the Soviet leaders would not attack. As the Soviet Union developed its own reconnaissance satellites, the Soviets saw mutual benefit in avoiding space warfare. A climate of mutual forbearance set in, which now serves both parties extremely well.

LESSONS LEARNED

By any measure, CORONA was a remarkable achievement. Its photographs provided the backbone of U.S. intelligence capability for 12 precarious years. It flew 145 times in that period, returning 167 film capsules with over 2 million feet of film. Its cameras covered a half-million square miles of denied territory—time after time. CORONA made an extraordinary contribution to world stability. It gave U.S. presidents poise and confidence when they were most needed. It guided U.S. national security policy during the worst years of the Cold War and eventually enabled arms control treaties to be negotiated and monitored with confidence—treaties that are now reducing nuclear and conventional arsenals dramatically. This first satellite reconnaissance system was truly a triumph of American technology.

Coming at the beginning of the space age, CORONA taught important lessons. It showed that one must clearly understand the objectives of a space mission before proceeding. CORONA's objective was to get broad area coverage at 25-foot resolution as quickly as possible. One must be sure that the technology is ready in such undertakings. Technology is constantly changing, and it is right to anticipate and build on that progress. On the other hand, if one is working on an urgent program, one cannot depend on a romantic view of technological progress. Finding the natural intersections of technology and national need is a difficult task. Only a few gifted people seem to do it well.

CORONA showed that using a small team of committed people is the right way to carry out such programs. All experience since then has shown that using accounting systems, business school approaches, and management slogans cannot substitute for a small team of intelligent and highly motivated people.

If CORONA teaches anything, it is to be courageous and persevere. One can only admire the Bissell-Ritland team. They launched relentlessly until they got it right. Their incredible gift to the nation was the result of vision and determination. The country owes them an enormous debt.

DWAYNE A. DAY

3 THE DEVELOPMENT AND IMPROVEMENT OF THE CORONA SATELLITE

In 1956 the U.S. Air Force formally started design and development work on a reconnaissance satellite at the Lockheed Missiles and Space Company. Although by then the Air Force had been studying the concept of reconnaissance satellites for a decade, work on the satellite proceeded slowly, due to a lack of funds as well as other projects having higher priority. Ballistic missile research was considered more important, and until an adequate ballistic missile was developed, a reconnaissance satellite would have no way of reaching orbit.

Sputnik, and President Eisenhower's intervention in the reconnaissance program in February 1958, would dramatically change this. Eisenhower directed that the CIA play a major role in the development of an "interim" satellite that would serve until the more ambitious Air Force satellite became available. It would operate under the cover story of a scientific and engineering program called Discoverer. A team of CIA and Air Force officials, together with engineers from industry, began work on this interim satellite. It was later said that the small number of people involved and their devotion to the program led to few turf battles.[1] Everyone involved considered the development of a reconnaissance satellite to be a top priority for the United States.

The basics of the CORONA program were established by early April 1958, but with a lot of details left to be determined. The CIA's Richard Bissell, who had overall responsibility for CORONA, scheduled the first CORONA spacecraft to be completed by June 1959. The interim program was planned to last only until June 1960. Bissell drafted a work statement for Lockheed on April 25 and began selecting supporting subcontractors.[2] General Electric would be responsible for the reentry vehicle. Itek would also be a subcontractor, but to lessen the blow to Fairchild (which had assumed that it would design and build the reconnaissance cameras), Bissell decided that Itek would define the basic camera concept, manufacture the lenses, and oversee camera design and fabrication, whereas Fairchild would design and manufacture the cameras. This was also necessary because Itek lacked the resources to manufacture its own cameras. The contractors began work on April 28 and submitted designs for first review on May 14. Designs were frozen on July 26, 1958.[3]

Because of the secret nature of the program, Lockheed engineer and program manager James Plummer and his handful of assistants did their initial work in a room in a small hotel away from the main Lockheed factory complex. Plummer was told by Lockheed management to run this program like Kelly Johnson ran the Skunk Works, which produced the U-2 aircraft. The Skunk Works was a unique production company with highly streamlined management, the ability to operate with a high degree of secrecy, and a distaste for paperwork. So the program managers decided to do most of the work off-site from the main Lockheed complex in Palo Alto. As part of the prime contract, Lockheed leased an unused advanced development facility from Hiller Aircraft in nearby East Palo Alto.[4] The facility, labeled "Advanced Projects," was closed to all Hiller personnel, who assumed that Lockheed was conducting classified helicopter work. Those who worked at the facility referred to it as "AP" or as CORONA's "Skunk Works."

The AP facility was to be used for the design and fabrication of the spacecraft nose structure, support system, and for payload integration. The cameras and Satellite Recovery Vehicles (SRVs) would come from manufacturers on the East Coast. Everything would be assembled and tested before it was shipped to the launch site for integration with the Agena upper stage and the Thor booster. The AP facility started out rather simply, but grew as the program expanded. While Bissell and John Parangosky, Deputy Chief of the CORONA Program Office Development Staff of the CIA, ran the program from Washington, Brigadier General Osmond Ritland and Lieutenant Colonel Lee Battle ran the program on the West Coast.[5] This new arrangement also relied upon streamlined management whereby program manager Jim Plummer and Lieutenant Colonel Battle reported *only* to Bissell and Ritland. This made decisions and accountability much easier than in other space programs of the period.

Due to the short life expected for the CORONA project, only twelve Thor boosters were procured; it was anticipated that the Air Force's WS-117L (now called SENTRY) reconnaissance satellite would replace CORONA rather quickly. But Bissell soon realized that the Air Force was not going to provide the boosters he had planned. He knew he needed four test launches and three launches for biomedical experiments in support of the CORONA cover story. The Advanced Research Projects Agency (ARPA), which felt that other military space missions also deserved support, refused to allocate DoD funds for these boosters. Bissell thus found himself having to go to a fiscally conservative president and ask for money to purchase nineteen launch vehicles not included in his earlier proposal. Although Eisenhower was not one to like these kinds of financial requests, he agreed to the additional funding.

Officially, Discoverer was supposed to be solely an engineering and scientific program. Five biomedical vehicles were scheduled to be built, two to carry mice and one to carry a primate, with the other two to be held in reserve in case the primate vehicle failed.[6] General Electric was responsible for developing the life-support systems for the mice and monkeys. A camera was placed between the monkey's legs during tests of the life-support system, so that the monkey's actions and emotional state could be observed. When the monkey's stomach got in the way, the engineers installed a mirror over its head and shifted the camera so that it picked up the monkey's reflection. In protest, the test monkey smeared feces all over the mirror. Even more serious, during ground tests of the primate life-support system, the monkeys kept dying.

LAUNCHES AND RECOVERIES

Photoreconnaissance satellites had to be in a polar orbit to maximize their coverage of the globe, which spun beneath the satellite as it traveled from pole to pole. The existing launch facilities at Cape Canaveral could not be used because the launch vehicles would overfly populated areas. Camp Cooke, located near California's Point Arguello, already selected as the launch site for the WS-117L program's Atlas launch vehicles, was perfect for CORONA's purposes. A rocket launched from Cooke to the south would not overfly any populated areas. Cooke was also the home of the 672nd Strategic Missile Squadron, which operated the Thor booster. In October 1958 Cooke was renamed Vandenberg Air Force Base.

One feature of Vandenberg presented problems for CORONA (as well as all other launches from the base). The Southern Pacific Railroad tracks run directly through the base. Because CORONA satellites had to pass over the Soviet Union in daylight as well as be recovered near Hawaii in daylight, they had to be launched from Vandenberg in the afternoon. The overall launch window was thus broken into several smaller windows dictated by the Southern Pacific schedule.[7] This was a safety concern, not a security concern—launch crews did not want to risk damaging a train if there was a launch failure.

Development of the CORONA payload and Agena upper stage continued throughout 1958.[8] The initial camera design consisted of a reciprocating panoramic camera that would sweep through 70 degrees of arc and then slide back to its initial position. The film would travel forward to a take-up reel inside a gold-plated and insulated bucket (which looked more like a round-bottomed kettle).

Requirements for the camera system conflicted. It had to be sturdy enough to withstand the high vibration of a launch, but it also had to be light enough to be carried by the extremely weight-limited Thor launch vehicle of the time. Once it was in orbit, the camera's lens/film relationship had to be maintained very precisely even though vehicle temperature would vary considerably as it traveled from hot sunlight to cold darkness every 45 minutes. All of this had to occur within an extreme vacuum, which was not a normal operating environment for any previous camera system.[9]

The Agena would serve as a second stage for the Thor, placing the payload in orbit and then supplying orbital power and stabilization. The Agena itself worked surprisingly well despite its complexity.[10] Once in orbit, the Agena would immediately yaw 180 degrees so that the SRV faced the rear. This minimized the amount of gas used for stabilization and protected the reentry vehicle from molecular heating, a poorly understood but worrisome phenomenon.

Tests began on the reentry vehicle's recovery system. Although the balloon program that contributed the basic camera design to CORONA had used an aircraft recovery system, that system had never been perfected or used extensively. The newly established 6593rd Test Squadron began training to make in-air snatches of payloads descending on parachutes. It operated out of Hickam Air Force Base in Hawaii, one of the bases attacked during the Pearl Harbor raid that had so galvanized the American intelligence program. Initial tests using various types of parachutes were disappointing. A different chute was also tested, again with disappointing results. The parachutes had high descent rates, making it difficult for planes to approach and snag the parachute lines. A new chute reduced the descent rate from 33 feet per second to 20 feet per second.[11] If a capsule was missed and landed in the water, it would float while its strobe light flashed and its radio beacon emitted a steady signal.[12] After one to three days, a plug made out of salt would dissolve and the capsule would fill with water and sink, thus preventing its recovery by an adversary.[13]

FIRST FLIGHT

CORONA/Discoverer preparations were conducted throughout California. The payload was prepared at AP, south of San Francisco, and shipped by truck to Vandenberg on the mid-state Pacific coast. The Agena was prepared at Lockheed's main facility in Palo Alto, and shipped to Vandenberg. Systems integration and other preparations were conducted at Air Force Ballistic Missile Division at Los Angeles Air Force Base in southern California. The final preparation of all components and overall vehicle integration and testing was done at Vandenberg. After being tested individually, the payload and the Agena would both be taken to the launch pad and mated to the Thor booster. Everything would be tested again before the launch vehicle's protective cover was removed and the rocket hydraulically lifted to a vertical position. Final checks would be made by the launch crew before liftoff.[14]

The plan was to prove all of the vehicle components quickly and sequentially. The first launch would test the Thor and Agena and the satellite control network. The second, third, and fourth flights would test the recovery sequence and related equipment, and would include the biomedical payloads. The fifth flight would include the first CORONA camera.

But the plan ran headlong into harsh reality. The first flight test of the Thor-Agena rocket combination (minus a payload and not assigned a number in the Discoverer series) was scheduled for January 21, 1959. During the last seconds of the countdown, a short circuit caused the explosive bolts connecting the Agena to the Thor to detonate. The small solid propellant ullage rockets, intended to fire just after the Agena's separation from the Thor in order to push the propellants to the back of the tanks, also fired on the ground. The thrust from the rockets caused some damage to the Thor, and the Agena was a total loss.[15] This test was commonly referred to as "Discoverer Zero," or "1019" after the number of the Agena vehicle.

A review conference held two days later at Lockheed quickly identified the short circuit as the result of poor systems integration testing. It also identified the changes needed to correct the problem with little alteration of the schedule. The 1019 mission served as an example of how *not* to conduct launch preparations, and "1019" quickly became a rallying cry and a warning for those concerned with launch operations. Extensive testing of the entire system became a long-lasting rule for anyone associated with military space programs.[16]

On February 28, 1959, Discoverer I was launched carrying only a light engineering payload. It was not heard from again, although several ground stations claimed to have tracked it. Based on these initial reports, and more than a little optimism, the Air Force issued a press release declaring that the satellite was in orbit. Air Force Major Frank Buzard, in charge of writing the mission report, was told to prove that the spacecraft went into orbit. Consequently he issued a report saying that the vehicle had indeed entered orbit. Later, he concluded that the Agena stage must have malfunctioned during the second half of its flight and that Discoverer I had probably landed somewhere near the South Pole.[17]

Discoverer II was launched on April 13. The capsule carried a small biomedical payload consisting of a "mechanical mouse." Two days later the Air Force announced that plans to recover the capsule near Hawaii had been canceled and that it would instead be recovered in the Arctic.

In reality, a human programming error had caused the capsule to be ejected early, and it had come down near the Spitzbergen Islands north of Norway.[18]

Although the islands belong to Norway, the Soviet Union had a lease to operate several mining facilities on them. The Air Force was less than accurate in its announcement because it did not want the Soviet Union to obtain the capsule. Its fiery descent and colorful parachute had supposedly been witnessed by Spitzbergen residents. The U.S. Air Force, with the aid of the Norwegian military, began an intensive search for the capsule. On April 22 the Air Force terminated its search, declaring the capsule lost. An explanation was contained in a memo from Nathan F. Twining, Chairman of the Joint Chiefs, to the Deputy Secretary of Defense: "From concentric circular tracks found in the snow at the suspected impact point and leading to one of the Soviet mining concessions on the island, we strongly suspect that the Soviets are in possession of the capsule." Twining's memo also mentioned reports of ground sightings of the capsule hanging from its parachute, and provided a list of Soviet mining facilities in the area, the contents of the capsule, and a draft of a possible diplomatic approach that could be used by the State Department to obtain the return of the capsule and its contents.[19] There were also rumors that a Soviet fishing trawler had suddenly left port right after the event. These rumors were not included in the memo.[20]

If an overture to the Soviets was made, it was unsuccessful, and the capsule was never recovered. From the capsule, the Soviets could have determined some preliminary information about the CORONA program, although what value that information was to them is questionable. The size of the capsule could have given them an idea of the diameter of the film cassettes carried inside later satellites, although that would have been of little use since a number of other factors determined the amount of film that CORONA could carry.[21]

Not everyone was convinced that the capsule had been recovered. Major Buzard initially felt that the capsule had come down on Spitzbergen, but later changed his mind and suggested that the vehicle probably came down in the water and sank, or never came down at all, suffering the same fate as later satellites that had problems with their retro-rocket de-spin system. There was no evidence on the ground to prove that Discoverer II had been recovered. "Witnesses" had been told exactly what to look for and may have merely repeated what they were told.[22]

Discoverer III was the first mission to carry animals—four black mice—as part of the cover story for the CORONA program. A launch attempt in late May was aborted when telemetry indicated a lack of mice activity. It was thought that the mice were asleep. When a technician banged on the side of the vehicle and failed to wake them, the launch was aborted. In fact, the mice had consumed the krylon coating on their cages and died.[23]

The second launch attempt a few days later was aborted when a humidity sensor in the capsule reported a 100% relative humidity level. The mice had

This unusual angle of the Discoverer VII preparing for launch from Vandenberg Air Force Base, California, on November 7, 1959, reveals the white protective cover at the top of the rocket. This provided cooling to the payload on the launch pad and also shielded its identity from outside observers. This was the fourth launch of the C camera (KH-1). However, this satellite failed to reach orbit when its Agena upper stage malfunctioned. (Photo courtesy of the U.S. Air Force)

urinated on the sensor, causing the faulty readings. The vehicle was dried out and readied for another launch attempted on June 3. During vehicle separation the Agena apparently fired straight down, plunging the Agena, payload, and mice into the Pacific Ocean. The mishap resulted in complaints from both a British humane association and several American newspapers, which were also incensed about the death of the monkey Able, who had perished after a joint Army-Navy suborbital flight in a Jupiter vehicle on May 29.[24]

Because of the urgent need to get intelligence information (Bissell, Ritland, and everyone else privy to the program knew that the U-2's days were numbered), the program directors chose to fly a camera at the earliest opportunity.

Thus, despite the elaborate cover story identifying Discoverer as having a biomedical payload, when the first camera became available before the life-support system could be made to work, no primate flights were ever conducted. The biomedical story did get publicity even if the more elaborate missions never flew, however, and helped to mask the true identity of the payloads.

Discoverer IV was launched on June 25, 1959. It was the first rocket to carry a reconnaissance payload, a C-model reciprocating panoramic camera; it was given the mission number 9001. The C camera was a direct descendant of the HYAC-1 camera developed for the WS-461L balloons. Unlike the HYAC camera, the film was located alongside the camera instead of above it. Designed by Itek and manufactured by Fairchild Camera and Instrument Company, it was an Tessar f/5 lens camera and had a focal length of 24 inches—twice that of the HYAC-1 camera.[25] It swung through an arc of 70 degrees—less than that of the HYAC-1 camera—and exposed a swath of film 2.10 inches wide.[26] It swung at a right angle to the direction of flight (the spacecraft faced rearward), so that at the center of its swing the camera faced straight down toward the earth.[27] The C camera had constant velocity image motion compensation (IMC)—the compensation for the movement of the image as the satellite traveled at five miles per second. This meant that the satellite could only fly at one altitude or the images would appear blurred.

Contractors faced a number of unforeseen problems. Because the camera oscillated back and forth during operation, it imparted undesirable motions to the spacecraft. To counteract this, the designers installed a momentum balance wheel that oscillated in opposition to the lens/scan arm assembly. Fairchild engineers chose to use a set of gears and a ball pushing on a spiral screw shaft to produce the necessary motion, but this resulted in significant vibration.[28] All problems were not of a technical nature either. Although Itek developed the lenses and Fairchild built the cameras, Itek did not have enough lens-grinding and -polishing machines or craftsmen to provide all the necessary lens elements. It had to subcontract this work to several precision optics firms.[29]

The C camera was designed to have 20–25-foot ground resolution, but at its best only achieved 35–40-foot ground resolution—which was still significantly better than what had been proposed for the earlier television-based and film-readout systems.[30] Discoverer IV never reached orbit to test the camera, because of another Agena failure. But there was no time to waste. The program's managers immediately began preparing for another launch.

"AUTUMN LEAVES"

Discoverer V, CORONA Mission 9002, also carried a camera but, unlike its immediate predecessor, *did* reach orbit on August 13. However, the temperature inside the vehicle was abnormally low and the camera failed on the first orbit. Telemetry on the ground indicated that no film ever made it into the return bucket, implying that it broke at some point during its path out of the film supply container. The reentry vehicle's retro-rocket fired it upward instead of downward and it ended up in an orbit with an apogee of 1,058 miles. The launch of Discoverer VI, CORONA Mission 9003, quickly followed on August 19. Discoverer VI's camera failed on its second orbit, probably also due to film breakage. The retro-rocket on its reentry vehicle apparently failed as well. Discoverer VII, CORONA Mission 9004, was launched on November 7; the Agena did not place it in orbit. Thirteen days later, Discoverer VIII, CORONA Mission 9005, was launched, but went into an eccentric orbit and once again no film went through the camera. Although the recovery vehicle ejected successfully, the capsule disappeared after separation and there was no way to know of its fate. CORONA managers had a Cold War to fight and could not take time to dwell on the failures.

At approximately the same time that the project engineers determined that the film in the satellites was breaking before going through the camera, ground tests in an atmospheric chamber yielded the same results. The acetate-based film degassed solvents in the vacuum of space and became brittle. It did not tear—it crumbled, like "autumn leaves," in the words of one of the Itek engineers. Since the same kind of film had operated without problems on the WS-461L balloons at the near-vacuum of 80,000 feet, no problems were envisioned in orbit. But the hard vacuum of space was a significantly different environment than the upper atmosphere, and the CORONA camera's film path, with its twists and turns (because of the need to accommodate the small diameter of the satellite), was much more complicated.

Eastman Kodak solved the problem. It obtained a license from DuPont to produce a polyester film base, which it coated with a high-resolution emulsion. This proved to be a very important technical feat. Polyester-based film was not unusual, but getting the high-resolution emulsion to bond to it required some creative chemistry on the part of Kodak.[31] This film also weighed only half as much as the acetate-based film. Kodak's high-resolution, lightweight films made aerial reconnaissance, and later space reconnaissance, possible.

WORKING UNDER CURSED STARS

There were other problems besides the film. Problems occurred in many different systems, but each one allowed engineers to learn something new about the system. At the same time, each failure also meant that the CORONA program was nowhere near operational. The "interim" program was rapidly depleting its initial supply of boosters and cameras. Bissell and Ritland concluded that CORONA needed greater quality control and more testing on the ground. They began a series of improvements to address the myriad problems.

The Agena had been developed for use on both the Thor and Atlas boosters. The latter was far more powerful and would be used for SAMOS (the renamed SENTRY) and the early warning MIDAS satellite. In order to conserve weight on CORONA, the Thor was allowed to burn until its fuel was exhausted instead of being shut down.[32] This, coupled with the low altitude of the CORONA orbits, required extreme precision at orbital injection. An angular error of ±1.1 degrees or a velocity deficit of 100 feet per second would result in the vehicle failing to achieve orbit. Only a bigger booster, which was not yet available, could solve this dilemma. In the meantime, technicians cut weight from the vehicle by using tin snips and files to cut away surplus metal. This solution was crude, but it worked.[33]

Another problem centered on the spin and de-spin rockets used to stabilize the recovery vehicle during reentry. They were designed to fire to "spin up" the SRV before retro-rocket burn, and then to fire again to stop the spin as the SRV reoriented itself according to aerodynamic forces. However, the propellant used in the tiny rockets was very unreliable. Project engineers had established an elaborate procedure for testing the rockets in batches, but defective rockets still made it through the testing phase and then failed in orbit. Fortunately, the solution was both elegant and simple—the two rockets were replaced by a cold-gas system that used a single nitrogen tank firing through two nozzles. This system went into operation with Discoverer XII, and the problem was solved.[34]

After the failure of Discoverer VIII in November 1959, a 2.5-month stand-down was ordered to correct the myriad problems. Once CORONA engineers felt confident that they understood the problem, launches resumed. On February 4, 1960, Discoverer IX, CORONA Mission 9006, was launched. This was the first mission to use the new polyester-based film, but it was never tested because the Agena did not achieve orbit. On February 19, Discoverer X, CORONA Mission 9007, was launched. This time the normally reliable Thor booster started to wobble soon after launch and the vehicle was destroyed. The payload was located by helicopter and retrieved by a crew who drove to the site by jeep.[35]

About this time discussions about canceling CORONA surfaced. Rumors circulated that the CIA simply was not capable of handling the job. Morale was at a low point. Air Force Colonel Paul Worthman, responsible for maintaining contacts between the CIA and the Air Force, reminded CORONA engineers that problems in such a pioneering and rushed program were inevitable. Bissell worked hard to squash the cancellation rumors. He urged everyone to push on.[36]

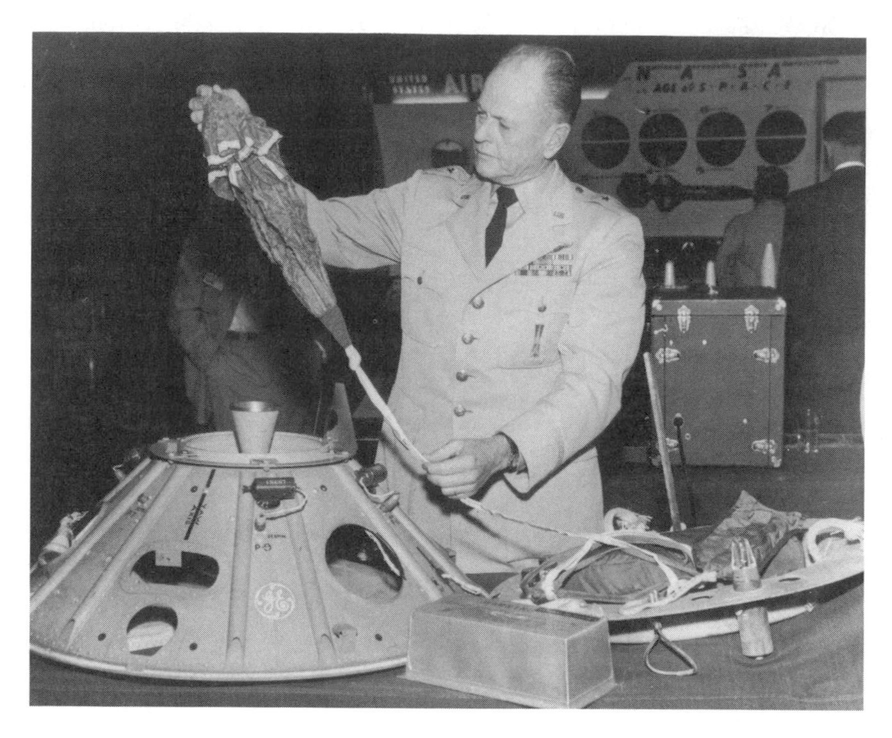

Brigadier General Osmond Ritland looking at the recovery package for the CORONA system. Note the small solid rocket motors around the circumference of the base cone of the reentry vehicle, used to spin it prior to retro-rocket burn. Reliability problems with the rocket motors led to their replacement with a cold-gas system. (Photo courtesy of the U.S. Air Force)

Discoverer XI, CORONA Mission 9008, was launched on April 15 and achieved orbit. This was the first mission to orbit carrying the new polyester-based film, and it was primarily because of the new film that the camera operated successfully. Unfortunately, the spin rockets apparently exploded during the reentry sequence and the capsule was not recovered.[37] Yet another stand-down was ordered until the problems could be worked out on the ground. This led to the incorporation of the cold-gas spin system previously mentioned.

Another problem addressed during the stand-down was the extreme difficulty of determining exactly what had gone wrong in orbit. Once a satellite was launched, only a limited amount of telemetry existed to indicate the nature of the problem. For instance, when film was breaking, the engineers only knew that it was not reaching the collection bucket. They did not know if this problem was due to the camera or the film; the film's disintegration in ground tests gave the necessary clue. Similarly, once the SRV separated from the Agena, there was no more telemetry. If the reentry vehicle did not come down to Earth, it could have been oriented in the wrong direction and fired up instead of down; or its small de-spin rockets could have exploded; or its engine could have not fired at all. All of these options had to be evaluated, taking extra time. The lack of good telemetry data was very frustrating to those attempting to make CORONA work.[38] Program managers decided to fly Discoverer XII as a heavily instrumented diagnostic flight, without a camera, but with the new cold-gas spin and de-spin system. Unfortunately, Discoverer XII, launched on June 29, 1960, failed to reach orbit when its Agena upper stage malfunctioned. By this time, however, the success of the camera on the previous flight and faith in the reliability of the cold-gas system led virtually everyone involved to be optimistic about the chances for success on the next mission.[39]

In the meantime, the urgency of the program had also increased. On May 1, 1960, the Soviet Union, using an SA-2 missile, shot down Francis Gary Powers in his U-2 spy plane over Siberia. U-2 missions came to an abrupt halt and the best source of data on the strategic capabilities of the Soviet Union disappeared. The U-2 program had been far more successful and lasted longer than anyone associated with it expected. Nevertheless, its loss came as a distinct shock to many who had come to rely on it so heavily. CORONA was now the last, best hope to fill the gap.

FIRST LIGHT

The Discoverer XIII launch on August 10, 1960, was a repeat of the previous instrumented test flight. This time the vehicle performed as planned. But there was a glitch in the recovery effort, as the recovery airplane headed in the wrong direction. Fortunately, a Navy ship, the *Haiti Victory,* retrieved the capsule, which was returned to Washington, D.C., and presented to President Eisenhower during a White House ceremony. The only payload in the capsule was an American flag, which was given to Eisenhower.[40] The ceremony and flag reinforced the fiction that Discoverer was solely a scientific and engineering program.[41] After this success, CORONA managers were ecstatic. In East Palo Alto, Lockheed employees celebrated with a party at a local hotel. They threw the program manager, James Plummer, into the swimming pool and jumped in after him. (See photographs in chapter 9.)

Discoverer XIV, CORONA Mission 9009, launched on August 18, was the first totally successful mission. It returned photographs of the Mys Schmidta airfield in the Soviet Union and its capsule was successfully snatched out of the air by a C-119 flown by Captain Harold Mitchell. Earlier missions had carried 10 pounds of film, instead of the planned full 40-pound capacity. Mission 9008

Members of the 6594th Squadron transporting a film-return bucket (contained inside the drum) aboard a C-130 Hercules aircraft. The 6594th was the premier Discoverer recovery unit. The drum was transferred to any available Air Force aircraft at Hawaii headed for California and then transported to the Advanced Projects facility, where the film was removed for transfer to the East Coast. (Photo courtesy of A. Roy Burks)

had carried 16 pounds. But Mission 9009 carried 20 pounds, all of which was successfully exposed and transferred to the reentry vehicle.

The recovery sequence involving an air catch above the Pacific was the one aspect of CORONA that most Air Force officers—particularly pilots—found hardest to accept. It just was not possible, they felt. In actuality, the air catch proved to be one of the easiest aspects of CORONA to perfect. Pilots practiced by catching parachutes and buckets dropped from high-flying U-2 aircraft over the desert. The C-119 Flying Boxcar was soon replaced by the larger and more capable C-130 Hercules. The challenges of making complex machinery work in the harsh environment of space were immense, but soon the air catch of returning payloads operated like clockwork and pilots even perfected night recoveries of illuminated space capsules.

After retrieval, the film followed a complex route to the customer, photo-interpreters in Washington. Aboard the C-119 or C-130 the bucket was loaded into a light-tight bag inside a steel drum and transported to Hawaii. There the drum was transferred to whichever Air Force plane was available at the time and flown to Moffett Field or Travis Air Force Base in California.[42] The drum was then taken to the Advanced Projects facility, where the bucket was removed. The film cassettes were flown to a processing center in the Northeast for development.[43] The original negative was developed and copied and two duplicate positives were produced for use by the photo-interpreters. The duplicate positives were flown to Andrews Air Force Base near Washington, D.C., and were delivered to the Photographic Interpretation Center (later NPIC) for analysis.[44]

Before the first mission returned its film, photo-interpreters working on U-2 photographs from earlier missions over the Soviet Union were told to put away their U-2 materials and await the arrival of new photographic imagery, which they all knew was going to be from a satellite. Many of them found the initial results from CORONA disappointing. CORONA returned more imagery of the Soviet Union than the 24 previous U-2 missions combined and revealed 64 new Soviet airfields and 26 new surface-to-air missile (SAM) sites. The resolution of the imagery was good enough to spot airfields and to look at vast areas of previously uncovered territory, but it was not as good as high-resolution U-2 photographs for hard intelligence gathering (such as counting actual bombers). After quickly exhausting the intelligence potential of the CORONA imagery, many of the photo-interpreters went back to their dated U-2 materials.[45] About 50 percent of the ground targets on the CORONA imagery were obscured by clouds, but several important targets were photographed.[46] The press release trumpeting the air catch of the SRV did not mention any of these facts, however.[47]

CHOOSING THE TARGETS AND EVALUATING THE IMAGES

In early 1959, Director of Central Intelligence Allen Dulles formed the Satellite Intelligence Requirements Committee (SIRC). The SIRC was to manage targeting for satellite programs independent of the Ad Hoc Requirements Committee (ARC), which established targets for the U-2 program. After the U-2 was shot down in May 1960, Dulles decided to replace both entities with a joint committee.[48]

On August 9, 1960, Dulles established the Committee on Overhead Reconnaissance (COMOR). COMOR was to

> coordinate the adaptation of priority foreign-intelligence objectives and requirements established by USIB [United States Intelligence Board], members of USIB, or other committees of USIB, to the capabilities of existing and potential overhead reconnaissance systems; and shall examine and make recommendations, as appropriate, on such related matters as dissemination and any special security controls required, but shall not undertake to provide operational guidance.[49]

In short, COMOR was in charge of setting targeting priorities for intelligence collection by CORONA.

The USIB, for instance, might establish a requirement for more information on Soviet nuclear facilities. While other members of the Intelligence Community would seek information through other channels (e.g., clandestine agent reports and signals intelligence intercepts), COMOR was responsible for seeing that relevant imagery was collected. COMOR would identify specific nuclear facilities to be imaged; this list would then go to an operations center (first at the CIA and later at the National Reconnaissance Office [NRO], where it was eventually known as the "Mushroom Factory"), which was responsible for outlining the basic operational parameters of a mission.[50] This small group would determine launch and recovery times, orbital paths, when to operate the cameras, and other details. This outline would then go to engineers at the Lockheed AP facility, where it was converted into specific commands for the spacecraft. CORONA's commands were stored on 35mm Mylar tape with holes cut into it at specific points. Five wire brushes ran along one side of the tape. When a brush encountered a hole, it would touch a metal contact on the other side of the tape, closing a circuit. With several brushes, a large number of circuit combinations were possible, and thus a large number of commands could be given to the spacecraft. The system proved quite reliable, but was also limited in flexibility at first, since all commands for the spacecraft were preset before launch.

Getting the spacecraft in orbit and photographing certain targets was important, but the photographs still had to be evaluated once they were returned to Earth. During the first U-2 flights, the CIA operated the Photo-Intelligence Division (PID). By 1958, the Army and Navy joined with the CIA (the Air Force wanted to conduct its own evaluations) and formed the Photographic Intelligence Center (PIC). In fall 1960, a major imagery review group that had evaluated the Intelligence Community's ability to conduct photo-interpretation recommended to President Eisenhower that the Air Force efforts be incorporated into the PIC. Eisenhower agreed with the recommendations. The National Photographic Interpretation Center (NPIC) was created in January 1961, with Arthur Lundahl as the first director.[51]

On August 18, 1960, the same day that the first film was successfully retrieved from space, James Q. Reber, chairman of COMOR, issued the "List of Highest

Priority Targets: USSR." This list included ICBM and IRBM bases and production complexes, submarine-launched ballistic missile and heavy bomber bases and manufacturing sites, and nuclear energy facilities—a total of thirty-two in all, complete with geographic coordinates. Most of the facilities were believed to be along major rail lines—which was one of the ways that U-2 targets were selected. Included in the list were the cities of Kirov, Danilov, Konosha, Murmansk, Kyshtym, Sverdlovsk, Mogilev, Leningrad, Gorkiy, and Sevastapol, to name a few.[52]

On this same day, a staffer to the Joint Chiefs, Colonel James E. Mahon, issued a memorandum calling for a review of scheduling of future CORONA launches with the idea of pushing the program into operational status as soon as possible. Mahon noted that the needs for reconnaissance and geodesy (the location of geographical points on the earth) information were both critical. He suggested that COMOR recommend that a CORONA mission using the newer C′ ("C Prime") camera system be launched as soon as possible to obtain reconnaissance of the whole of the Soviet Union, particularly focusing on targets set forth in a revised May 25 "List of Highest Priority Targets." Mahon stated that as soon as that mission was accomplished, a mapping mission using the CORONA spacecraft but a different camera, known as ARGON, should be launched to fulfill the geodesy requirements. In other words, CORONA would identify the targets and ARGON would pinpoint their location.[53]

The need for accurate geodetic information was very important from a strategic warfighting point of view. A bomber crew could be expected to search for its target and then attack it. But ICBMs have an inherent inaccuracy during their flight that is measured in terms of circular error probable (CEP), which, simply stated, indicates how near to their target half of the missiles fired will fall. If the exact location of the target is unknown, errors compound upon each other and there is less assurance that an ICBM will actually hit its target. ARGON was needed to remove a large part of this uncertainty by photographing vast swaths of land so that accurate geodetic measurements could be taken.

ARGON had been approved as an Army-sponsored independent mapping project a year earlier, on July 21, 1959.[54] It replaced an Air Force SAMOS reconnaissance camera system known as E-4. ARGON was handled under the auspices of the CORONA program out of the need for security and a concern that it might compete for launches. ARGON employed a 3-inch focal-length still camera with 5-inch-wide film and a resolution of approximately 460 feet. Its images covered an area of 300 by 300 nautical miles.[55]

BUILDING BETTER SPACECRAFT

New versions of CORONA were also in development. Discoverer XVI, CORONA Mission 9011, was launched in October 1960 with the first C′ camera, but failed to reach orbit. The C′ was also designed and manufactured by Fairchild, under Itek supervision. The C′ differed from the C by having variable image motion compensation so that different orbits could be flown.[56] Discoverer XVII was launched a month later but suffered a payload malfunction. However, Discoverer XVIII, CORONA Mission 9013, launched on December 7, 1960, worked perfectly.

Discoverer XVI was also the first launch using the Agena B second-stage booster. The new booster failed during this launch but was successful during the following launch. The Agena B represented a significant improvement over the A version. Not only did it have bigger fuel tanks—effectively doubling its fuel capacity (and hence increasing its payload and safety margins) over early Agena versions—but it also possessed a restart capability.[57] Restarting a space vehicle in orbit represented a difficult challenge, but the only significant additional equipment required for the Agena, other than the increased fuel load, was a second starter charge and additional ullage rockets that fired to push the fuel to the rear of the tanks so that it could feed into the engines.[58]

On February 17, 1961, the first ARGON satellite was launched, with the designation CORONA Mission 9014A (the A stood for ARGON). But the orbital programmer and the camera failed and there was no recovery of film from this mission. The next CORONA mission after that also failed. The next two ARGONs after that failed. The following CORONA was a success. The one following that was only a partial success after the camera stopped midway

Imagery of a Soviet airfield returned from CORONA Mission 9017 on June 16, 1961. This satellite used a C′ (KH-2) camera, had a maximum resolution of about 25 feet, and proved to be particularly successful; it provided the details and locations of several key Soviet strategic facilities. (Photo courtesy of the National Reconnaissance Office)

through the flight. Then there was another ARGON failure (due to a rare Thor malfunction), followed by another CORONA malfunction.

On August 30, 1961, the first C‴ ("C Triple Prime") camera was launched. (There was no C″ version.) The C‴ was the first Petzval f/3.5 lens camera, the first camera both designed and built by Itek, and the first camera to include significant improvements to reduce vibration. The C‴ differed from its predecessors by having the lower section of the camera containing the heavy lenses rotate a full 360 degrees while the scan arm at top rocked (or oscillated) back and forth through a 70-degree arc.

Robert Hopkins of the University of Rochester had suggested using the Petzval f/3.5 lens on CORONA. He realized that the field curvature in a Petzval lens was not as important when it was being used in a panoramic camera. The Itek-designed Tessar f/5.0 lens used on the earlier C and C′ cameras did not have sufficient light-gathering power to allow use of a slower film that could record finer detail. By switching to the Petzval lens, the designers could also switch to a slower film. Typical film speeds for household cameras are in the ASA 200–ASA 400 range. CORONA ultimately used films with speeds of ASA 2 to ASA 8, depending on the processing.[59] Inserting a field flattener in the focal plane created diffraction-limited resolution, allowing the camera to achieve 180–200 lines per millimeter—much better than the C and C′ cameras at 100–120 lines per millimeter. For the earlier cameras, enlargements could be made up to 20 times the original image size. The new camera allowed enlargements up to 40 times the original image size.[60]

The Petzval design was large. It used three 7-inch diameter elements at the front and two 5.5-inch diameter elements at the rear. These were all encased in a 22-inch-long cylindrical cell. The lens rotated around an axis about 10 inches from the front of the cell. In addition to the five lenses in the cell, a sixth lens segment, acting as the field-flattener, was located .25 inches in front of the film. Collectively, the six lenses weighed about 20 pounds.[61]

The C‴ camera was also modified to reduce the effect of thermal differentials on its components. Its controls were made more reliable, and the method of metering film and achieving and maintaining camera focus were also improved. Adjusting the IMC depending upon the orbit allowed for lower orbits and higher resolution. Furthermore, timing pulses, which determined the scan velocities and IMC for each frame, were marked in the image area rather than the border area of the film. Resolution improved from 25 feet for the C′ to 12–25 feet for the C‴. Finally, 39 pounds of film could now be carried.[62]

This mission, number 9023, was a success. Of the sixteen CORONA and ARGON missions launched in 1961, seven CORONAs returned film (all of the ARGON missions were failures). Gradually, problems were being solved. Also in 1960 and 1961, two CORONA missions had been launched without cameras

After a launch failure, wreckage from the launch vehicle and the classified payload would be returned to a classified facility for painstaking examination in order to determine the cause of the failure. (Photo courtesy of A. Roy Burks)

or SRVs. They were instead equipped with radiometric sensors for the MIDAS early warning satellite program. Both missions were successful.[63]

MURAL AND MAPPING

In addition to problem-solving, CORONA engineers were working on more ambitious plans. On August 9, 1961, a contract was awarded to Itek for a more capable camera system. This contract was made retroactive to March 20—a procedure that appears to have been common during the CORONA program; improvements were often begun before they had been officially approved. This new camera system was known as the "MURAL" or "M" camera, and it filled an obvious requirement from an intelligence standpoint—the need for stereoscopic imagery.

MURAL actually consisted of two C‴ cameras. One was mounted so that it was tilted slightly forward and the other so it was slightly backward, so that their fields of view overlapped. The forward-looking camera (the camera in the "front" of the spacecraft, which was actually faced backward along the flight path) photographed a swath at a 15-degree angle. About a half-dozen frames later, the rear-looking camera also photographed the same swath at a 15-degree angle, but from a different direction. When the two film swaths were properly aligned in a stereo-microscope, the image would appear to be three-dimensional and interpreters could obtain accurate height data on ground objects.[64] For the Itek design team, when compared to the engineering problems encountered with the earlier cameras, MURAL proved to be relatively easy—almost fun—to design.[65] In practice, the cameras worked perfectly together most of the time, but on the rare occasion when one camera failed, useful images could still be obtained from the remaining camera.[66]

The two cameras received their film from a dual film supply cassette at the rear of the payload. The two film strips traveled forward and were then bent at a 90-degree angle when they reached their respective cameras. Once exposed, they were then bent again and sent forward to film take-up cassettes inside the reentry vehicle buckets.

The last C‴ camera was launched aboard Discoverer XXXVII, CORONA Mission 9030, in January 1962, but the Agena failed to orbit. A month and a half later, the first MURAL camera was launched on Mission 9031.[67] This launch, on February 27, was publicly known as Discoverer XXXVIII. It was the last Discoverer launch. Those in charge of CORONA had realized that the cover story had worn thin; there were now simply too many launches for this to be considered an experimental, scientific program. Thereafter, all launches would simply be listed as classified Air Force launches under the designation "Program 162."[68]

Mission 9031, in addition to the MURAL camera, also included a small Index camera with a focal length of 1.5 inches that took a small-scale photograph to be used for matching the panoramic swaths to the terrain. A few missions later a nearly identical Stellar camera was added which, in combination with the horizon cameras, provided a much more precise indication of vehicle attitude.[69]

The panoramic camera sacrificed geometric relationships in order to maximize image detail. The true geographic location of an object on the panoramic film frame could be positioned to only a few miles, using knowledge of the orbital track and altitude, the time that the photograph was taken, and the attitude of the spacecraft. But this was not precise enough for targeting purposes and hence the government asked Itek to develop a camera that would photograph a large area on each frame to permit the panoramic photographs to be precisely positioned within a geographic framework.

The Index camera, as it was known, was actually a commercial Hasselblad camera modified for use in space. The German-manufactured Hasselblad is a very high quality camera frequently used by professional photographers, but it

A KH-4 CORONA payload during vibration testing at Lockheed's Advanced Projects facility. The KH-4 had only a single film-return bucket. This version of the spacecraft was bigger than the early version (shown in the introduction) due primarily to the addition of a second camera within the cylindrical section. But it was not as big as the KH-4A shown in chapter 2, which included a second film-return bucket below the first.

was not designed for use in a weightless vacuum, where lubricants and sealants could evaporate and condense on a cold lens surface, thereby fogging the image. It was also not designed to carry the large amount of film that would be required for heavy on-orbit use. The camera required some redesign. Early versions of the Index camera had a slight distortion problem. The camera was redesigned with a new shutter manufactured by a commercial firm lacking any security clearance; thus the firm could not be told the purpose for which the shutters were to be used. The manufacturer of the shutter used talcum powder as a shutter blade lubricant, but this was not acceptable for space use. Itek engineers suggested the use of whale oil instead. Because whale oil cost $11,000 per gallon, the manufacturer balked. This meant that Itek engineers had to accept the new shutters with talcum, disassemble them, clean and lubricate them with whale oil, and reassemble them. Only about a drop of oil was needed for each shutter, meaning that Itek was left with a lot of unused whale oil.[70]

Panoramic images could be used for mapping purposes in combination with the dedicated Index camera. Because they were curved, one inch at the center of the photograph represented a different ground distance than at the ends of the photograph, which meant that the images were not suitable for mapping purposes. Itek manufactured optical rectifiers—essentially a photographic enlarger that reversed the original process (in which the "flat" earth was imaged on the cylindrical film surface). In the optical rectifier, the film was held in cylindrical form (just like in the camera) and a light behind it projected an image via a lens onto a flat sheet of photographic paper which represented the flat earth (the rectifier also compensated for the earth's curvature). The print that was produced removed the distortion inherent in the film process; one inch on the print represented the same ground distance, no matter where it was located on the image.[71]

Before CORONA, geographic positions of many locations within denied territory were known to no better than an accuracy of 30 miles. Due to the development of cartographic cameras, by the end of CORONA they were known to within 400 feet in the horizontal plane and 300 feet vertically. Many points were known to within 150 feet in the horizontal plane.

Sometime before the first MURAL camera was flown, officials at the NRO, which was created in 1961 to assume overall charge of American reconnaissance efforts, designated all reconnaissance satellites with the code name KEYHOLE and a numerical designator. The MURAL system was labeled KH-4. The designation was retroactively extended to earlier cameras: the C-camera missions were labeled KH-1, the C′ missions KH-2, and the C‴ missions KH-3. ARGON was labeled KH-5.

Sixteen KH-4 missions were launched in 1962, with only two failures. Three ARGON missions were successfully launched during this same period. Satellite reconnaissance had become an operational intelligence method, well integrated with the intelligence process overall.

CORONA PLAGUES CORONA

Film from early CORONA flights occasionally returned faded or marked with spectacular branch-like patterns. This phenomenon, common to many different reconnaissance cameras, was known as "corona." Although the problem had been somewhat sporadic and inconsistent in earlier CORONA missions, CORONA Mission 9040, launched in July 1962, returned with extensive film damage due to corona and radiation fogging.

The problem was not new to the optical engineers who had designed the CORONA cameras, although the extent ofthe problem was unexpected. Corona occurred when the film dried out in the vacuum of space after unwinding from the film-supply cassette on its way to the camera. Static electricity built up on the moving parts of the camera, particularly the rubber film rollers, and the resulting sparks affected the film. Although the cause was well understood, finding a solution was far more problematic. A group of Itek environmental test-lab people, after careful trial-and-error testing, replaced the material in the film rollers with a new material that did not build up any static electricity.[72] The rollers were ground to size and steamed with water in a pressure cooker, and then mounted in the engineering model of the camera and tested in a vacuum chamber. If they produced no corona marking, they were tested in a flight camera two more times. If they failed any of these tests, they were rejected and refurbished. The acceptance rate was about 10 percent. Frank Madden, who was then in charge of the camera at Itek, had to procure precisely machined film rollers from subcontractors who could not be told what they were for, a factor that limited their ability to respond to the problem.[73]

The corona problem never went away entirely. Since it was most noticeable on the first frames when the film had been exposed to the vacuum of space for some time, the cameras were usually started a few frames before reaching the target so that only unaffected film was used over the target. After changes were made to the film rollers, and careful precautions were taken, corona ceased to be a major problem.[74]

While the corona problem required a special committee and a lot of effort to find a solution, CORONA program engineers continued to work on correcting other flaws and making improvements to the system. After each mission, project engineers made minor adjustments to equipment being prepared for future flights. More important, top officials from the major contractors (Lockheed, Itek, and General Electric) met with representatives from NPIC, the CIA, and the Air Force. First they discussed the mission's performance, evaluating the quality of the images and any problems (such as film stretching or flaking). After discussing possible fixes, they proposed improvements to the overall system.

CORONA satellites were procured in "batches" of six spacecraft each. The general rule was that no design changes would be made inside a batch; adjustments would be allowed only when starting a new batch. But ideas for improvements came so rapidly—particularly from the photo-interpreters at NPIC who were using the imagery and running head-first into its limitations—that the CORONA engineers began inserting changes within the batches.[75] Horizon and stellar cameras were added to provide accurate positioning data on the spacecraft. A stereo camera system was placed on the KH-4, an Index camera essentially took the place of the KH-5 ARGON, and fiducia markers (small, evenly spaced lines) were incorporated for taking measurements from the film.

SAMOS

As CORONA evolved, unplanned developments had an impact on the program. The original plan called for the use of only nineteen Thor rockets, but

A float test of a CORONA film-return bucket off the coast of California (probably near Santa Barbara) in the mid-1960s. The buckets had a salt plug that after a day would completely dissolve, and the bucket would fill with water and sink, assuring that it would not be retrieved by an adversary. (Photo courtesy of A. Roy Burks)

rocket—and camera—quotas had been reached by late 1960, with only one successful film-return mission. The ARGON mapping mission, with its own requirement for Thor boosters, had been added to the program, taking over a mission previously handled by the Air Force. The Air Force's SAMOS program, which had been expected to replace the "interim" CORONA program, was running into delays and cost overruns. Concern about this situation, as well as a desire to ensure that the program was focused on *national* intelligence needs as opposed to *Air Force* intelligence needs, led to the creation of the Office of Satellite and Missile Systems in the Pentagon in August 1960.[76] This office did not report through the normal chain of command at the Air Force, but to upper levels of the Pentagon.[77]

Since CORONA was showing the first signs of success, and SAMOS was delayed while its management was overhauled, the CORONA program was extended and improvements made to its camera system. This kept the CIA in the satellite reconnaissance business. Although CORONA was still not intended to be a permanent program, it would continue until SAMOS could replace it.

By 1960, SAMOS consisted of four separate reconnaissance projects: SAMOS E-1, a readout satellite that would broadcast its data to the ground via radio and with 100-foot resolution of ground targets (it was primarily a test of the equipment); SAMOS E-2, a readout high-magnification surveillance camera with a design resolution of 20 feet; SAMOS E-5, a film-return high-magnification satellite with a design resolution of 5 feet; and SAMOS E-6 (sometimes referred to as SAMOS 201), a film-return high-magnification satellite with a design resolution of 8 feet.[78] SAMOS E-6 was to be manufactured by a different contractor than the other satellites to create a redundant satellite production capability.[79] SAMOS also consisted of dedicated ELINT (electronic intelligence) satellites, as well as attached payloads and smaller "ferret" satellites.[80]

The Air Force's SAMOS program included a film-recovery program as well as the film-readout system because top Air Force officials eventually recognized that film return, even though slower than film readout, had its own benefits—specifically higher resolution. So the Air Force decided to develop a film-return satellite independent of CORONA before CORONA became operational. But in June 1959, ARPA deferred developing a recoverable capsule for the SAMOS program; it withdrew support for it in the 1960 budget, halting the Air Force's plan for a film-return satellite of its own.[81] This action greatly displeased Lieutenant General Bernard Schriever, who was then head of the Air Research and Development Command. Schriever sent a letter to Air Force Chief of Staff General Thomas D. White: "Should the ARPA decline to continue the recovery program . . . it is recommended that the Air Force immediately support this urgent development."[82] Responding for White, Deputy Chief of Staff General Curtis LeMay replied, "I am completely sympathetic with your point of view and have taken action through Secretarial channels to restate the Air Force requirement to the director of ARPA and request reconsideration of its support in FY 60."[83]

ARPA had been established in February 1958 to oversee the United States' military space programs. Although it had some oversight of the "Discoverer" aspect of the CORONA program, it did not exert much real authority on the covert reconnaissance program; CORONA managers strove to keep ARPA in the dark about what they were doing. In other areas of the military space program, however, ARPA exerted a great deal of control, such as over SAMOS and MIDAS. ARPA was not popular with the Air Force, which viewed it as infringing on its decisionmaking authority on space issues. ARPA was not popular with the other services for many of the same reasons, although they also feared that the Air Force would predominate if ARPA was simply eliminated.

In April 1959, the Chief of Naval Operations urged the Joint Chiefs of Staff (JCS) to create a single military space agency. The Army, rapidly losing its space program to NASA, agreed. The Air Force Chief of Staff objected and stated that this would remove weapons systems from the unified commands. Within three months White House and DoD officials were exploring the idea of a separate military space agency, tentatively called the Defense Astronautical Agency, which would report directly to the JCS; command would rotate among the services.[84]

That September, the Secretary of Defense rejected the idea of a separate military space agency. He returned military space responsibility to the separate services from ARPA. Booster development went to the Air Force; payload development went to the Army, Navy, and Air Force based upon competence and primary interest.[85] ARPA now became simply a technology development agency for the Department of Defense. With ARPA out of the way, the Air Force began full-scale development of its own film-recovery system, the SAMOS E-5, followed in 1960 by a second film-recovery program, SAMOS E-6.

The first SAMOS film-readout satellite launch took place on October 11, 1960. The associated press release contained considerable information on the satellite, including the fact that it carried a photographic payload. But the satellite never reached orbit. A second SAMOS launch took place in January 1961; this time the press release omitted a reference to the photographic payload. The satellite successfully reached orbit and its film-readout system began operating. But by this time, SAMOS E-2 was only a technology demonstrator and not intended to become operational.

The Air Force pursued the film-readout version of SAMOS as a means of obtaining battlefield reconnaissance. For this mission, timeliness—getting the images back quickly so that military actions could be based on them—was more important than resolution or area coverage. SAMOS E-2 was intended to fulfill this requirement and to stay in orbit much longer than the short-lived CORONA satellites. One of the limitations on such a system was that the satellite had to carry enough film for its entire mission. Another, more important limitation was the amount of data that could be transmitted over the radio link to the ground. As the satellite passed over the ground station with its nose pointed at the earth, it would slowly rotate in order to keep the side with the radio antenna pointed at the ground station to extend the data transmission time.[86] Because of the need for security, the ground stations had to be located within the United States. There was just no way to transmit a great amount of data from the satellite before the satellite moved out of range of the ground station.[87]

The Air Force began flying the SAMOS E-5 satellites in 1961. Equipped with solar panels and a more sophisticated panoramic camera than CORONA, the E-5 was expected to produce higher-resolution coverage of a smaller area. The first E-5, launched in November, failed to reach orbit; a subsequent mission in December reached orbit but did not return successfully. A third mission in March 1962 also reached orbit, but failed to return its film. SAMOS E-5 was soon canceled, and the remaining four or five cameras built for the program were placed in a classified government warehouse.

SAMOS E-6 conducted five launches beginning in April 1962. Like all reconnaissance systems, it suffered startup problems and did not successfully return film until November 1962. By this time its capabilities were not really any better than the KH-4 with its MURAL cameras. SAMOS E-6 was eventually canceled like its predecessors. The "intended successor" to CORONA had proved to be a bust.

FAILED PHOENIX: KH-6 LANYARD

In April 1960, the last successful U-2 mission over the Soviet Union photographed a prototype ABM facility at Sary Shagan. Two years later, CORONA returned a similar image from near Tallinn in Estonia. CIA officials were worried that this was an indication of Soviet deployment of an operational ABM system. But they needed more information to determine if the missiles were intended to shoot down ballistic missiles or aircraft; the CORONA images were not adequate.

In 1962, the director of the CIA's newly created Directorate of Research was Dr. Herbert "Pete" Scoville. After learning of the unusual facilities at Tallinn, Scoville asked Undersecretary of the Air Force and Director of the National Reconnaissance Office Joseph V. Charyk for help. Charyk agreed to pull the SAMOS E-5 cameras out of storage and develop a "crash" program to place them aboard rockets and launch them into orbit.[88] Their target would be the suspected Tallinn ABM facility. This effort was envisioned at best as a temporary program.

Two Lockheed engineers from the CORONA program, Bob Leeper and Bill Cotrell, traveled to the warehouse to evaluate the condition of the cameras for operational use. The E-5 had a 66-inch focal length, which required that it not be mounted through the diameter of the satellite (like the CORONA cameras), but lengthwise. The camera was a single-lens cell that obtained stereoscopic coverage by swinging a mirror through a 30-degree angle. Leeper and Cotrell found the cameras in good condition.[89] Soon thereafter, the E-5 cameras were transported to Itek for refurbishment. The E-5 camera and the spacecraft that Lockheed built for it were named the KH-6 and given the code name LANYARD. They were adapted for use with an Agena D upper stage and an uprated Thor booster.[90]

Five LANYARD spacecraft were produced. Of these, only three were launched. The first launch took place on March 18, 1963, atop the second "Thrust Augmented Thor," or TAT, Agena D rocket. The TAT was a Thor rocket with three small solid rocket boosters strapped alongside. The Agena D was a more powerful version of the basic Agena vehicle with bigger fuel tanks.

The Agena D was developed by Lockheed due to the fact that the company had practically custom-built several different versions of the Agena B for military and civilian payloads. The Agena D standardized the basic spacecraft control bus. Systems unique to each payload could be plugged into the forward section of the Agena, thereby increasing its versatility. Its reliability and precise pointing capability were valuable assets for reconnaissance missions. A few years later, the Agena would be adapted for tests with NASA's Gemini spacecraft and become the standard upper stage for U.S. military space missions for the next 24 years, with over 360 launches.

The first LANYARD mission, given the designation Mission 8001, also carried a Lockheed-built P-11 scientific subsatellite. It did not reach orbit due to a problem with the Agena D. Mission 8002, launched in May, reached orbit, but its Agena D malfunctioned in flight. Mission 8003, launched at the end of July, also reached orbit, but its camera failed after 32 hours. The camera suffered from lens-focusing problems due to thermal failures. The thermal effects were corrected in cameras still on the ground.[91] Intended to produce imagery with two-feet resolution, LANYARD reportedly achieved no better than six feet resolution—which was still better than CORONA imagery at the time. The stopgap LANYARD program was soon canceled. After many years of debate, it was determined that the Tallinn site was not really an ABM facility. Despite the lack of success with LANYARD itself, it did test several systems for later use: the Agena D saw use on KH-4 Mission 9038 in June 1962 and then was adopted for CORONA as the standard upper stage to be combined with the TAT.

THE "INTERIM" PROGRAM BECOMES PERMANENT

The next major upgrade of the CORONA came with the introduction of the KH-4A, first launched on August 24, 1963.[92] Because of the substantial differences in this vehicle, KH-4A launches were given a new mission launch sequence. Thus, even though it was the fifty-eighth launch in the CORONA/ARGON series, it was labeled Mission 1001.

The primary modification to the basic KH-4 design was the addition of a second SRV behind the first. This was commonly referred to as a "two-bucket system." For the KH-4A the film strips had to be sent to two take-up spools in two separate reentry vehicles. Engineers designed a transfer roller into the hub of the take-up spool in the second satellite recovery vehicle (SRV-2), which was closest to the cameras. This allowed the film to wind around the transfer roller and exit the SRV. They also installed an intermediate roller assembly, which was mounted to a vehicle bulkhead just in front of the forward panoramic camera.

As the film strip left the camera, it went into the intermediate roller assembly and then into and out of SRV-2 by means of the transfer roller in the hub of SRV-2. The film then went back to the intermediate roller assembly, which redirected it past SRV-2, through a film-cutter and into SRV-1, where it was connected to the take-up reel.

When the two film take-up reels in SRV-1 were full, or when ground controllers felt that the SRV contained critical information, the controllers sent a command to the spacecraft to cut the film. The forward cut ends of the two film strips (one per camera) would be wrapped up into SRV-1 and then the door seal closed. The rear cut ends of the film would be wrapped onto the SRV-2 spools.[93] SRV-1 would then detach from the spacecraft, leaving SRV-2 ready to accept film from the cameras.

The spacecraft could be stored in orbit in a passive, or "zombie," mode for up to 21 days, although missions usually lasted no more than 15 days.[94] This change in mission profile required a major redesign of the command and control mechanisms on the spacecraft. It also allowed a significant increase in the film load. While KH-1 and KH-2 had flown with no more than 20 pounds of film, the KH-4A eventually carried up to 160 pounds of film for its two buckets. The camera system, known as the J-1, was essentially identical to the MURAL camera. The only changes required were to double the capacity of the film-supply cassette and to strengthen the main support plates to accommodate the increased weight.[95] Resolution was slightly improved to 9 to 25 feet, although it was occasionally as good as 7 feet.[96]

Several other improvements were incorporated into the KH-4A during its lifetime. An independent, self-contained subsystem known as LIFEBOAT was built into the Agena. LIFEBOAT could be activated to recover the SRV in the event of an Agena power failure. The more powerful Thorad rocket was introduced with Mission 1036 and became standard five missions later, in May 1967. The Thorad was a TAT version of the Thor with more fuel. Small "drag make-up units" were also added to the Agena. Because CORONA operated with a low perigee to increase its resolution, it suffered from atmospheric drag, which threatened to de-orbit it prematurely. The drag make-up units were small rockets that could be fired individually at desired intervals to increase the vehicle's velocity and boost it back up to the proper orbit.[97]

The KH-4A had some troubles. Initially it experienced a severe heat problem, later corrected by improving shielding, altering the paint pattern on the vehicle, and reducing the thermal sensitivity of the camera. On the first mission, the master horizon camera remained open for extended periods of time, seriously fogging the panoramic photography. Its current inverter also failed in mid-flight, making it impossible to recover the second SRV.

During the peak of the CORONA program, there was a standing requirement to have six spacecraft at the AP facility ready to launch within 30 days.

CORONA was never launched that often, however, and usually maintained a pace of approximately one satellite launch per month.[98] The KH-4A returned a truly massive amount of intelligence data. KH-1 through KH-4 had returned a total of 285,472 feet of film; KH-4A alone returned 1,293,025 feet.[99] This flood of data revolutionized the field of intelligence analysis in general and photo-interpretation in particular.

STRUGGLE FOR CONTROL OF CORONA

At the time the KH-4A was introduced in 1963, the CIA was still in charge of the CORONA payload. It managed the CORONA program from the Operations Division of the Office of Special Activities under the Directorate of Research, headed by Deputy Director (Research) Pete Scoville.[100] The Operations Division had five branches: the Control Center, OXCART Branch, CORONA/ARGON Branch, IDEALIST Branch, and the Weather Branch. OXCART was the code name for the supersonic A-12 reconnaissance airplane. IDEALIST was the code name for the U-2. OXCART, an exciting and challenging program, was occupying the time of many of those in the Operations Division. By contrast, CORONA had become routine, almost boring.

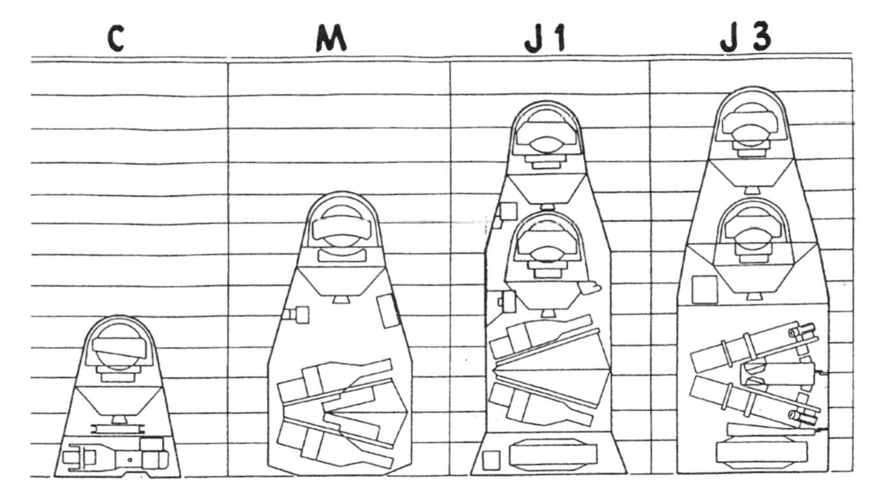

Schematic and size comparison of the CORONA variants. The "C" model, or KH-1 (KH-2 and KH-3 were similar), was only five feet in length and achieved its first successful mission in 1960. The "M" ("MURAL" or KH-4) first flew in 1962 and contained a second camera. The "J-1" (KH-4A) contained a second reentry vehicle and was first launched in 1963. The "J-3" (KH-4B) was the ultimate CORONA design and incorporated the Constant Rotator camera. It first flew in 1967 and was the last CORONA to be launched.

The National Reconnaissance Office, formally established in September 1961, was intended to oversee all U.S. reconnaissance satellite programs. By 1963 the CIA had largely lost interest in CORONA; the agency had a single representative at AP involved with managing CORONA, and he was detailed to the agency from the Air Force.[101] The CIA was not performing any other work on satellite reconnaissance at the time. By late winter of 1963, Pete Scoville was frustrated with both his undermanned directorate and his attempts to preserve the agency's presence in overhead reconnaissance.[102]

As a result of the CIA's inattention, the Air Force assumed more and more responsibility for CORONA within the framework of the highly classified NRO, while continuing to conduct a myriad of other military space missions outside of it, like the development of communications satellites, the Dyna-Soar spaceplane, and cooperative work with NASA on the Gemini program. Many in the Air Force felt that the service should simply take over CORONA completely. The fact that the CIA barely participated at the AP facility meant that this would not be a big step.

Secretary of Defense Robert McNamara was opposed to duplicating effort within the military bureaucracy and sought to eliminate it wherever he could. But he was also opposed to duplicating effort within the entire government, and frequently suggested eliminating programs in NASA or elsewhere because they paralleled others underway within the Department of Defense. This belief led him to pursue the TFX fighter plane in 1961–62, to propose that the Air Force completely take over the Gemini program from NASA in late 1962, and to eventually cancel the Air Force's Dyna-Soar spaceplane in favor of the NASA-developed Gemini manned spacecraft in 1963.[103] In this environment, it is not surprising that soon many within the Office of the Secretary of Defense were attempting to eliminate the CIA's involvement in satellite reconnaissance.[104]

In August 1963, Dr. Albert D. "Bud" Wheelon became head of the newly created Directorate of Science and Technology. Wheelon was a relatively young man with a Ph.D. in physics, who had agreed in 1962 to work for the CIA in Washington at the behest of Director of Central Intelligence (DCI) John McCone, Pete Scoville, and President Kennedy's science advisor, Jerome Wiesner. He agreed to take a job with the CIA for no more than three years, so that he could eventually return to a home on the West Coast and a career in private industry. Wheelon took over the directorate after Pete Scoville's resignation in March, but only with strong preconditions. He argued to McCone that reconnaissance was the most important mission that the CIA could perform. Reconnaissance, he felt, was a more powerful and useful "force multiplier" than even the hydrogen bomb, for reconnaissance allowed a president not only to fight a war, but, more important, to avoid it. The CIA could serve an important role as a "competing" satellite intelligence center to the Air Force.[105]

If Wheelon was going to assume Scoville's job, he wanted McCone's assurance that he would have the DCI's full support in the reconnaissance field. McCone assured him he would have it. Wheelon asked for, and received, far more authority over scientific and technical resources at the CIA than his predecessor, building the Directorate of Science and Technology into a major directorate at the CIA. In addition, he argued successfully for the creation of a higher pay scale with which to attract young scientists and engineers to the agency, particularly for reconnaissance work. While building his directorate, Wheelon angered many within the agency, as well as within the Air Force, but he established a powerful and impressive development and engineering organization.[106]

Wheelon soon revived the CIA's lagging interest in satellite reconnaissance and started three new intelligence satellite initiatives that remain classified.[107] He found that his plans conflicted with the wishes of Director of the National Reconnaissance Office Brockway McMillan and Deputy Secretary of Defense Roswell Gilpatric, both of whom were attempting to push the CIA out of the intelligence satellite field in order to consolidate it under Air Force control within the NRO. Wheelon, with McCone's support, fought these battles and gradually won most of them.[108] (See chapter 2 by Wheelon and chapter 6 on the National Reconnaissance Office.)

Although these skirmishes generally concerned other satellite programs and whether the CIA would retain a role in the satellite intelligence field, the conflict spread to CORONA.[109] Communications between the CIA and the Air Force became very poor, to the point where those responsible for Agena operations as well as overall systems integration for CORONA launches complained that they could not get information from those at the AP facility. AP told them that they did not have a "need to know."[110] In late 1964, apparently after a technical dispute over the spacecraft, the Air Force made a concerted effort to take over the CORONA program, culminating in McMillan suddenly reassigning the Air Force officer detailed to the CIA to oversee payload integration at the AP facility. The CIA immediately sent an agency employee out to California to take over that function.[111] During the course of about a year, as the war over funding and control of CORONA was being fought, the main contractor, Lockheed Missiles and Space, did not get paid for the payloads.[112] The firm absorbed the costs of CORONA while the two sides worked things out. Eventually Lockheed was paid for its work.

The relationship between the Air Force and the CIA, especially at Advanced Projects, improved, particularly after Al Flax took over as director of the NRO in 1965. From that point on, there was always significant CIA representation at Advanced Projects, and the agency did not let its attention to CORONA lag again.[113] The Air Force and the Department of Defense relented in their opposition to a CIA role in satellite reconnaissance, and the agency continued its work on CORONA and other programs. Despite this frequently bitter and raucous fight, actual CORONA activities never suffered, and the two organizations launched and operated the satellites successfully.[114]

KH-4B AND THE CONSTANT ROTATOR

The last KH-4A satellite, Mission 1051, flew in May 1969, but its replacement was already in operation, having been first launched almost two years before. In the spring of 1965 CORONA engineers began looking at new modifications to the spacecraft design. Representatives from the CIA, Air Force, Lockheed, General Electric, and Itek met to discuss the potential for improving the basic design. They established a series of design goals for a new camera for the spacecraft. These were:

Improve photographic performance by removing reciprocating scan arms and reducing vibration from moving components.

Improve the velocity-over-height (Image Motion Compensation) match to reduce image smearing.

Improve photographic scale by accommodating proper camera cycling rates to enable altitudes as low as 80 nautical miles (the J-1 camera operating altitude was 100 nautical miles).

Eliminate camera failures due to film pulling out of guide rails (an occasional problem with the J-1 camera system).

Improve exposure control through variable slit selection to get greater performance at low sun angles (as opposed to the J-1, which had only a single exposure for the entire mission).

Allow handling of different film types and split film loads through addition of an in-flight changeable filter and a film change detector.

Enable use of ultrathin base film, yielding a 50% increase in area coverage with no increase in weight.[115]

The C‴ camera had separated the heavy part of the camera—the lenses—from the scan arm, which contained the aperture that exposed the film.[116] The "lens cell" rotated a full 360 degrees while the scan arm rocked back and forth through a 70-degree arc. As the lens cell completed its rotation it would connect with the scan arm and then slide through 70 degrees of arc while exposing the film. This design eliminated the high vibration of the original C camera, but still suffered from vibration when the scan arm moved back and forth and came to a full stop at the end of each scan. The MURAL and J-1 cameras were essentially two C‴ cameras connected together, which doubled the scan arm vibration. Vibration "smeared" the image on the film enough to lower resolution from the best that could otherwise be achieved with the cameras.

To solve this problem, engineers designed a panoramic camera that connected the lens cell and the scan arm and placed them in a drum that rotated a full 360 degrees. This was known as the Constant Rotator, or CR, and it was possible only because of the increased diameter of the payload that came with increased performance from the Thor rocket. Film was still exposed during a 70-degree angular segment of the drum's rotation, but the vibration due to stopping and reversing the scan arms was eliminated. This dramatically improved ground resolution and added greater versatility in use. Each camera had two changeable filters and four changeable exposure slits that would allow different films and lighting conditions to be exploited. The new camera system was designated the J-3 (there was no J-2). The proposed ultrathin base film was abandoned due to variable imagery quality and failures in ground tests, but all other aspects of the camera proved spectacularly successful.[117]

Changes were made to the spacecraft as well. The earlier Index and Stellar-Index cameras, designed and built by Itek, performed quite well on the KH-4 and KH-4A and were used until the advent of the KH-4B system. But the Index camera had a 1.5-inch focal length, which was less than ideal for cartography work. It was replaced by a 3-inch camera nearly identical to that flown on ARGON, but with a dual-looking stellar instrument for star-sightings.[118]

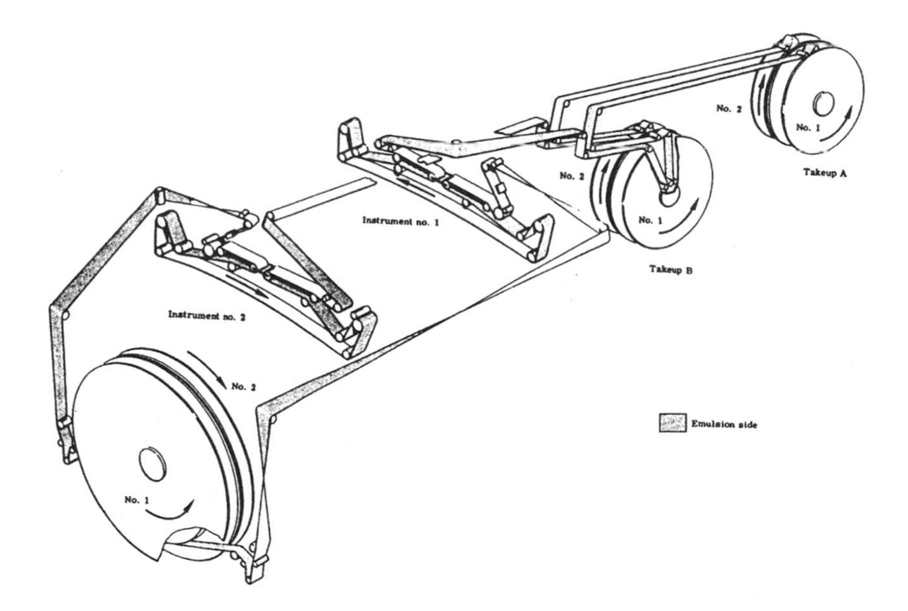

The film path for the KH-4B spacecraft. The film-supply cassette is on the left.

ARGON had not been very successful, but the new Dual Improved Stellar Index Camera (DISIC) worked quite well.

The J-3 camera worked perfectly from the start. The first KH-4B mission with this new camera was launched on September 15, 1967, and given the designation Mission 1101. Resolution with the new camera improved to as good as six feet (compared to the best possible resolution obtained under laboratory conditions, which was four feet). Because of the flexibility of the camera settings, the satellite could be taken down to altitudes as low as 80 nautical miles.[119]

In addition to better resolution, variable exposures and filters on the cameras allowed special color and high-speed, high-resolution black-and-white films to be carried.[120] A CORONA J-3 Ad Hoc Committee was convened by the director of the NRO on December 4, 1967, and began formal meetings in February 1968 to evaluate the potential of these new film types and techniques during a series of five flight tests. The committee was specifically interested in the intelligence uses of such film for new purposes, such as the detection of camouflage. Infrared film was flown on Mission 1104 and used to photograph the Vandenberg AFB area. Color film was flown on Mission 1105 and 1108.[121] The committee recommended further testing and also recommended the establishment of a special subcommittee of the Committee on Imagery Requirements and Exploitation (COMIREX, which had replaced COMOR on July 1, 1967) to evaluate the utility of color satellite photography.[122]

By 1970, the CIA had concluded that color film's decreased resolution made it relatively useless for intelligence purposes—black-and-white film had a resolution approximately twice as good as high-resolution color film.[123] But color film could be useful for mineral resources exploration. CIA thus offered a subcontract to a geology firm to evaluate the use of color imagery for mineral resources exploration. The study resulted in a report, "Appraisal of Geologic Value for Mineral Resources Exploration," which evaluated a film known as SO-242 flown on Mission 1108 in December 1969. It concluded that the color film had limited military-oriented intelligence utility, but could be of use to the nonmilitary Intelligence Community, such as the Offices of Basic and Geographic Intelligence and Economic Research at the CIA.[124]

NASA began its Earth Resources Technology Satellite program in 1969 and launched the first spacecraft in July 1972. The name was changed to Landsat before the second satellite's launch in January 1975.[125] Landsat used a light-sensitive electronic sensor that was far more advanced than the film-return cameras on the KH-4B, but had much lower resolution (due both to technology and political restrictions). The role that the CORONA J-3 Ad Hoc Committee and the KH-4B test flights played in early civilian earth resources policy and technology development is still largely unknown.

Various satellite reconnaissance programs contributed greatly to other, nonmilitary space programs. CORONA lenses found their way into a camera system used on the last Apollo missions. At least part of the SAMOS system was used for the Lunar Orbiter, which searched for Apollo landing sites. In addition, some of these systems and lenses were utilized on other reconnaissance platforms, such as the U-2 and SR-71.[126]

CORONA COMES TO AN END

By 1969, the CIA was no longer managing the procurement of CORONA equipment; it had turned over its responsibilities to the Air Force. Lockheed had also moved production from the AP facility to its main complex in Sunnyvale. The agency was busy working on other programs, including a CORONA replacement. The CIA began looking at a follow-on for CORONA in 1963 when CIA Deputy Director for Science and Technology Wheelon asked a group headed by Sidney Drell of Stanford to investigate whether CORONA could be upgraded to provide significantly higher resolution (on the order of two feet) for much vaster amounts of territory. The Drell group quickly determined that CORONA was limited in the degree to which it could be improved. Since the cameras ran through the diameter of the satellite, and since the satellite could be made only so wide, CORONA could be improved only so much before a new system was needed. The group recommended the development of a replacement. The CIA began this program in the mid-sixties and it first flew on June 15, 1971.[127]

There were still three more KH-4Bs in storage however. Charlie Murphy, who worked in the NRO's "Mushroom Factory" that was responsible for planning CORONA missions, heard a rumor that the last KH-4B spacecraft was going to be preserved for a museum. He thought that such a valuable piece of equipment should not go unused, and recommended to DNRO John McLucas that the remaining CORONA satellite be launched.[128] The data was still useful and the hardware was paid for. McLucas agreed, and the final KH-4B mission, Mission 1117, lifted off May 25, 1972. This was the last CORONA built. Even qualification models were refurbished and flown; by the time the program was ended there were no spare satellites.

Using the last film-return bucket returned from CORONA, along with the developmental model of the J-3 camera and photographs from the various missions, in 1972 a classified museum display of a CORONA spacecraft was set up inside the headquarters of the National Photographic Interpretation Center in Washington, D.C. At the dedication ceremony for the museum display, DCI Richard Helms discussed the importance of satellite reconnaissance and expressed the hope that the display could some day be turned over to the Smithsonian Institution.[129] Efforts at declassifying the CORONA program reached their final

The Thrust Augmented Thor launch vehicle used to launch later CORONA satellites. The three solid rocket motors at the base (one is obscured by the rocket itself) provided additional thrust. The protective shroud at the very top of the rocket is not inflated at this time. It was connected to an air conditioner and kept the payload cool. It also concealed the payload from outside observers and Soviet reconnaissance satellites. (Photo courtesy of the U.S. Air Force)

states before being vetoed at high level. It was not until February 1995 that the KH-4B model was displayed at a ceremony at the CIA headquarters in Langley, Virginia. The model is now on display in the National Air and Space Museum.[130]

CONCLUSION

CORONA was a classic example of Cold War military, industry, CIA, and think-tank cooperation. Satellite reconnaissance was conceived and nurtured by RAND, managed by the Air Force, adopted by the CIA, and made a reality by a combination of CIA–Air Force–contractor personnel, all working in consort to achieve spectacular results.

CORONA is as much testimony to the urgency of intelligence needs and the commitment of political leaders as to the hard work of the CORONA Pioneers. The program continued even after a string of failures and mis-starts. It was intended only as an interim solution to the need for intelligence and yet lasted 12 years. CORONA involved 145 launches, of which over 120 could be considered partially or totally successful.[131] The CIA's involvement in space-based intelligence collection was only supposed to be temporary; CORONA's success secured it a permanent place in the national security space program.

But while CORONA was created in part because of President Eisenhower's distrust of the Air Force, it also would not have been possible without Air Force involvement. CORONA relied heavily on equipment and techniques developed by the Air Force—mostly for programs not associated with space. The Air Force developed and managed the highly successful Thor launch vehicle which made CORONA possible, turning a ballistic missile into a successful space launcher. The Air Force was also responsible for the ambitious and difficult Agena upper stage, whose three-axis stabilization offered many advantages over simple spin stabilization. In addition, the aerial recovery technique for CORONA was a result of the Air Force's balloon reconnaissance program and relied exclusively on Air Force personnel and equipment. Even the reentry vehicles owed more to the Air Force than the CIA, since they were a direct outgrowth of ICBM technology.

In many ways the shape of the program was determined by the environment of the time and the interests of the organizations involved. The CORONA photoreconnaissance satellite series was developed to satisfy an urgent Cold War need for imagery of the Soviet Union and other denied areas. The CIA was more interested in hard intelligence about Soviet capabilities and intentions than the Air Force, which was more interested in locating targets to attack. These requirements were not simply arbitrary ones, but represented the distinctly different missions of the CIA and the Air Force.

Despite a difficult and discouraging beginning, CORONA was ultimately a tremendous success and proved the value of space-based reconnaissance for enhancing knowledge of a dangerous enemy. CORONA and the people who built and operated it deserve much of the credit for winning the Cold War.

R. CARGILL HALL

4
POSTWAR STRATEGIC RECONNAISSANCE AND THE GENESIS OF CORONA

The concept of strategic reconnaissance that emerged in the first few years after World War II differed sharply from accepted wartime convention. The Army Air Forces' (AAF) doctrine associated strategic reconnaissance with the well-established intelligence functions of identifying and pinpointing vital targets for aerial bombardment, with identifying any antiair threats that might impede the striking force, and with bomb damage assessment. With the advent of atomic weapons, however, a few young officers and civilian scientists and engineers, especially those associated with the AAF's Aeronautical Photographic Laboratory at Wright Field, Dayton, Ohio, began to think of strategic reconnaissance in quite different terms, as an intelligence tool that could be used to provide advance warning of a surprise attack. At first termed "pre-D-day photography," then "pre-hostilities reconnaissance," a redefined strategic reconnaissance mission called for acquiring reliable intelligence about the economic and military activities and resources of a potential foreign adversary through periodic, high-altitude overflight in peacetime.[1]

In the remarkably short span of 15 years (1945–60), as the Cold War increasingly shaped national perceptions and actions, this concept was embraced by American political leaders and made into national policy. Secretly, at the direction of two presidents, resources would be allocated and technology developed to put this policy into practice. First, military aircraft would be used for peacetime overflight of potential foreign adversaries. Next, specially designed, unarmored, high-flying reconnaissance balloons and aircraft would be so employed. Finally, by 1960, automated satellites would gather intelligence concerning activities on Earth while operating silently in outer space. The clandestine reconnaissance plans, national policy, and machines created in this period would provide far more diverse and reliable kinds of intelligence than that providing advance indications and warning of surprise attack. Collectively, they also made possible the sizing of the U.S. military establishment to meet actual threats rather than imagined ones, enabled arms control treaties with verification, and helped maintain a delicate peace between the Cold War's international protagonists.

A CONCEPT FORMED

The first person to articulate the concept of overhead strategic reconnaissance was Lieutenant Colonel Richard S. Leghorn, wartime commander of the 30th Photographic Reconnaissance Squadron in the European Theater. He had graduated from the Massachusetts Institute of Technology (MIT) in 1939 with a bachelor's degree in physics and a reserve commission in the Army Ordnance Corps. While still a college student, Leghorn served as an officer in the Sigma Chi fraternity, where he became well acquainted with its faculty "den mother" and MIT graduate, the 32-year-old editor of MIT's *Technology Review,* James R. Killian Jr. With a degree in English and without scientific credentials, Killian nonetheless possessed an extraordinary knack for facilitating collaborative efforts among scientists and nonscientists, and among students and faculty alike. Killian and Leghorn's paths would cross again many times in the years to come.

Richard S. Leghorn, whom many call the father of American strategic reconnaissance, 1955. (Photo courtesy R. Cargill Hall)

Working at Eastman Kodak in December 1939, Leghorn met Major George W. Goddard, commander of the Army Air Corps' small Aeronautical Photographic Laboratory at Wright Field, near Dayton, Ohio. Determined that his active military service would be in aerial photography, Leghorn offered to work for Goddard at Wright Field if the major could arrange to transfer his commission to the Air Corps. In early 1940, Goddard did just that. Shortly after Germany invaded Denmark, Norway, the Low Countries, and France in the spring of 1940, Leghorn learned he now held a commission in the Army Air Corps. At year's end he obtained an active duty assignment at the Aeronautical Photographic Laboratory.[2]

Second Lieutenant Richard Leghorn arrived at Wright Field in March 1941; there he worked on improving methods of assessing and exploiting data from aerial photographs. At Goddard's laboratory, Leghorn became acquainted with two individuals whose interests and careers also would become intertwined with his in the years afterward: Amrom H. Katz, a recently hired civilian physicist from Milwaukee, Wisconsin, who tested and compared the performance of camera shutter systems; and Lieutenant Walter J. Levison, a physicist from City College of New York, who tested and characterized light-sensitive materials and development processes in the laboratory's sensitometry unit.

Major Goddard also recruited for wartime services astronomer James G. Baker at Harvard University. German-made lenses were no longer available; Baker visited Goddard in December 1940 to discuss the situation. With the permission of his superiors at Harvard College Observatory, after January 1941 Baker designed and perfected lenses for a number of Air Corps reconnaissance cameras. During World War II, the Harvard Optical Laboratory, which Baker formed for this purpose, became a major component of the National Defense Research Committee's (NDRC) Division 16, which operated under the wartime Office of Scientific Research and Development.[3] The K-22 aerial camera was one of the most significant of Baker's contributions to aerial photography. This camera featured a 40-inch focal length f/5.0 distortionless telephoto lens that adjusted automatically to the pressure and temperature variations inherent in high-altitude aerial photography. By war's end, James Baker was widely recognized as America's preeminent lens designer for aerial reconnaissance cameras, a reputation he maintained with even more remarkable contributions for much of the rest of the century.[4]

Many of the men from Goddard's Aeronautical Photographic Laboratory, together with Duncan E. Macdonald, a physicist from Boston University, worked together in 1946 for the AAF, filming the two atomic tests of Project CROSSROADS at Bikini Atoll in the Pacific. Traveling between Roswell, New Mexico, where the photographic unit was formed, and Kwajalein Atoll, where it was stationed for the tests, Leghorn read a summary report of the *United States Strategic*

Bombing Survey (Europe), issued on September 30, 1945. This document assessed the effectiveness of and lessons learned from the strategic bombing campaign against Nazi Germany and the reconnaissance that supported it. A few key points impressed Leghorn. "In the field of strategic [target] intelligence," the report stated, the United States needed "more accurate information, *especially before* and during the early phases of the war." The report urged improving coordination among intelligence services in the collection and evaluation of information, and suggested involving civilian scientists in peacetime as well as in wartime. Underscoring these and other points, the report concluded: "The combination of the atomic bomb with remote-control projectiles of ocean-spanning range stands as a possibility which is awesome and frightful to contemplate."[5]

In addition to the frightening possibility of a future sneak atomic attack on the United States, what struck Leghorn most forcibly was the identified need for pre-hostilities targeting reconnaissance, separate and distinct from tactical reconnaissance during wartime. That distinction matched his own experiences in pre-D-day photographic missions flown from Great Britain and in subsequent combat reconnaissance conducted on the Continent. But in a world where more than one state possessed atomic weapons, securing strategic targeting data was insufficient to ensure survival. Leghorn also realized that the United States had

James G. Baker, director of the Harvard Observatory Army Air Forces Optical Project, adjusting a 40-inch focal length aerial camera developed for the service, ca. 1944. Baker was one of the pioneers of American reconnaissance cameras. (Photo courtesy of James G. Baker)

to have reliable indications and warning of force levels and enemy capabilities for a surprise attack *in advance,* if it was to avoid an atomic disaster. This information, Leghorn was sure, could only be acquired through periodic, extremely high altitude aerial reconnaissance overflight of potential adversaries using new aircraft especially designed for this task.

During the long hours between sorties on Kwajalein, he explained and elaborated upon this proposition to anyone who would listen, and especially to Duncan Macdonald, who was among the first converts to his cause. The CROSSROADS atomic tests that the AAF photographic unit filmed in July 1946, in particular the underwater test on July 25 that tossed the Japanese heavy cruiser *Nagato* 400 yards like a bathtub toy, also seared themselves in his memory and committed him to selling "pre-D-day photography," as he then termed it, to the AAF, to America's political leaders, and to the public.[6]

Leghorn's first opportunity to do so came a few months later in December 1946 when Macdonald invited him to speak at the dedication of Boston University's Optical Research Laboratory. Back in 1943, Colonel George Goddard had secured funding to build a proper optical workshop for James Baker and relocate his AAF lens design efforts from the basement confines of the Harvard College Observatory. The new brick structure, built on the grounds of Harvard University with reinforced footings that eliminated any vibrations from traffic on nearby roads, was completed in late 1944 and, as an element of the wartime NDRC, was dedicated as the Harvard Optical Laboratory. In this

The American "Baker" atomic bomb test at Bikini Atoll on July 25, 1946. This test was a seminal event for American strategic reconnaissance. Not only did it bring together many of the top photographic and reconnaissance experts at the time, but it also impressed upon them the incredible destructiveness of nuclear weapons.

well-equipped facility, Baker and his assistant Duncan Macdonald had finished designing the 60-inch f/5.0 lens and began work on an ultra-high-precision (K-30) 100-inch f/10 system using a 9 × 18-inch film format. The Perkin-Elmer Company in Glenbrook, Connecticut, fabricated all of Baker's lenses for the AAF. By war's end, the laboratory employed thirty-six optics scientists and fifty technicians. In September 1945, however, when Colonel Goddard visited Cambridge, Harlow Shapley, chief of astronomy at Harvard, advised him that the university would return to peacetime operations immediately. That meant severing all ties with military research. James Baker would return to the university full-time and the Harvard Optical Laboratory staff would be disbanded.[7]

Desperate to save his team of optical designers and technicians for national security research, a shaken Goddard appealed to Daniel L. Marsh, president of nearby Boston University (BU), to house the optical laboratory staff in three temporary wartime structures built by the federal government on the BU campus. The AAF would continue to fund the laboratory's operation. If Marsh and Chester M. Alter, dean of the BU graduate school, agreed, Duncan Macdonald, BU physics professor and Baker's wartime assistant, would serve as director of the new Boston University Optical Research Laboratory (BUORL).[8] In February 1946, Marsh and Alter did agree. After moving out those in residence in the buildings, re-equipping them, and transferring the Harvard staff, BUORL was formally dedicated on December 13, 1946, with Duncan Macdonald as director. The luminaries that attended the dedication signified the importance accorded the optics laboratory both locally and nationally. They represented the colleges and universities along the Charles River, virtually every major photographic and optical firm in the country,[9] and the military services. AAF leaders in attendance included Major General Curtis E. LeMay, Deputy Chief of Air Staff for Research and Development; Brigadier General Alden R. Crawford, Assistant Chief of Air Staff for Materiel; and Major General Laurence C. Craigie, Chief, Engineering Division, Air Technical Services Command.[10]

At the BUORL dedication Leghorn outlined the requirements for and impediments to achieving strategic overflight reconnaissance. A world in which more than one country possessed atomic weapons, Leghorn asserted, was a world that would "demand . . . aerial reconnaissance prior to the outbreak of hostilities." In that world, he continued,

> military intelligence becomes the most important guardian of our national security. The nature of atomic warfare is such that once attacks are launched against us, it will be extremely difficult, if not impossible, to recover from them and counterattack successfully. Therefore, it obviously becomes essential that we have prior knowledge of the possibility of an attack, for defensive action against it must be taken before it is launched. Military intelligence is the agency for providing this information, and our

national security rests upon its effectiveness, next to a sound international political structure.

Overhead reconnaissance conducted with cameras in daylight and, if they could be developed, radars at night and in overcast conditions would produce the core of this intelligence, particularly "in the case of potential enemies of a totalitarian, police-state nature where the acquisition of information by the older methods of military intelligence is more successfully blocked." Nonetheless, Leghorn had to concede, unauthorized overflight of a foreign state in peacetime was denied by treaty law and

would be considered an act of military aggression. It is unfortunate that . . . peacetime spying is considered a normal function [while] . . . aerial reconnaissance—which is simply another method of spying—is given more weight as an act of military aggression. Unless thinking on this subject is changed, reconnaissance flights will not be . . . performed in peacetime without permission of the nation state over which the flight is to be made.

Leghorn thought it highly unlikely that such overflight permission would ever be granted. Consequently, he concluded, to ensure national security the United States would have to devise means for overflight reconnaissance that could not be detected:

The accomplishment of this objective is not as technically difficult as it might at first appear. Extremely long-range aircraft capable of flying at very high altitudes are currently on the drawing boards, and in some cases prototypes have been constructed. Effective means of camouflaging them at high altitudes against visual observations are well known. It is not inconceivable to think that means of preventing telltale reflections of other electromagnetic wave lengths, particularly of radar frequency, can be developed. With such a tool at hand, information can be secured of a potential enemy's mining of radioactive materials and his plants—necessarily large—for the production of fissionable products, as well as a variety of other essential data.[11]

The kind of world that Leghorn described for members of the audience, however, did not yet exist in 1946. The United States alone possessed atomic weapons. The Army Air Forces preferred to modify multiengined combat aircraft for reconnaissance and to avoid spending scarce funds on single-purpose aircraft tailored to that mission alone. Finally and most discouraging, even "with such a tool at hand," international law proscribed unauthorized flight in the airspace of another state. Unless political and legal strictures against reconnaissance overflights were changed, any American leader ordering them could trigger a serious international incident, and might provoke another war.

A LEGAL OBSTACLE AND A POLITICAL WATERSHED

International air navigation treaties negotiated in the wake of World War I featured one cardinal principle: each state claimed for itself exclusive sovereignty over the airspace above its territory and territorial waters. Just as no land vehicle might enter another state without clearance at the border, no foreign aircraft might enter the airspace of another state in peacetime without prior permission. States' representatives agreed that the welfare and safety of each nation was no greater than its command of the air overhead. Article 1 in the first convention of 1919 made the point explicitly:

> The High Contracting Powers recognize that every Power has complete and exclusive sovereignty over the air space above its territory.
>
> For the purpose of the present convention the territory of a State shall be understood as including the national territory, both that of the mother country and of the colonies, and the territorial waters adjacent therein.[12]

Subsequent treaties on air commerce that U.S. presidents signed and the Senate ratified contained similar introductory articles. In terms of application, moreover, all international treaties to which the United States was a contracting party, under the Supremacy Clause of the Constitution (Article VI, Section 2), became the law of the land. Whether or not he had read these international instruments, it was this elemental aspect of air law to which Richard Leghorn referred in his 1946 BUORL address. Among the military members who heard him, no general officer could deny the primacy of civilian leadership or the rule of law. Only a president of the United States, as a head of state, might counter the strictures of international law—either by withdrawing his country from a treaty under the terms provided for it, or, for some overriding reasons of state, by expressly violating its canon and the law of nations.

In the months after the December 1946 dedication, the Cold War began in earnest. In an attempt to detect Soviet military preparations, the AAF in late 1946 began flying RB-29 reconnaissance aircraft along the USSR's northern borders nearest the United States. Equipped with cameras of limited focal length (36-inch or less), these aircraft flew within a few miles of the Soviet coastline to obtain oblique photography of Soviet territory. The two regions of greatest concern were the Chukotskiy Peninsula directly across the Bering Strait from Alaska and the Kola Peninsula north of Leningrad on the Barents Sea, which contained the port of Murmansk.[13] The Soviet Union, however, claimed sovereignty over territorial waters and the airspace above it within twelve miles of its coast, not the three miles recognized by the United States. To avoid any incidents, the Department of State instructed the AAF to observe the twelve-mile limit.[14]

Even so, the Soviets protested American "bombers" violating the airspace above their territorial waters. The State Department responded with aerial approach restrictions that eventually reached a distance of 40 miles.[15] A 40-mile limit, of course, precluded any useful aerial photographic reconnaissance missions. But after the Soviet Union detonated an atomic device in August 1949[16] and communist forces swept to victory in China in October, the stand-off restriction was reduced to 20 miles and, in the face of other provocations, the entire issue was reopened.

Soviet detonation of an atomic weapon well in advance of U.S. expectations greatly concerned American leaders. Curtis E. LeMay, by then commander of the Strategic Air Command (SAC), recommended that the United States employ pre-hostilities strategic overflight reconnaissance to detect Soviet preparation for a surprise attack and, more important, should adopt a preemptive war policy.[17] At the prodding of MIT's George Valley, the Air Force Scientific Advisory Board undertook an assessment of nearly nonexistent U.S. air defenses with the intention of identifying ways for improving them.[18] Soviet action in Europe had already prompted the Truman Doctrine (economic and military assistance that helped Greece and Turkey repel communist insurgents), the Marshall Plan (economic assistance that helped rebuild the war-ravaged Western European infrastructure), and the North Atlantic Treaty Organization (NATO). Now, at the president's direction, the National Security Council (NSC) began a sweeping evaluation of U.S. military preparedness and of the nation's objectives, strategy, policies, and programs for national security with respect to the USSR.[19]

A few months later, on June 25, 1950, North Korean forces launched a surprise attack on the Republic of South Korea, and in November Chinese forces entered that conflict in force. The sequence and pace of these events, coupled with available intelligence of Soviet forces in Eastern Europe, prompted American political and military leaders to believe that their Soviet counterparts might launch an attack against Western Europe, possibly accompanied by a surprise aerial attack on the United States. On December 16 President Truman issued a Proclamation of National Emergency, called numerous National Guard units to active duty, and signed an executive order creating an Office of Defense Mobilization to control all executive branch mobilization activities including production, procurement, manpower, and transportation. On December 19 the president advised the nation that General of the Army Dwight D. Eisenhower would return to active duty as Supreme Commander, Allied Powers in Europe (SACEUR), in charge of NATO forces. While the JCS assessed existing war plans and alerted American commanders to the possibility of a global war,[20] they also considered how aerial overflight reconnaissance might determine with greater certainty Soviet preparations for an atomic surprise attack.

The JCS chairman, General of the Army Omar N. Bradley, ordered a reassessment of aerial reconnaissance policy and then placed the issue before the president.[21] Reliable information on the status of Soviet aerial forces in those regions of Eastern Siberia closest to the United States could be acquired only through reconnaissance overflight. In the Korean conflict, the USSR supported North Korea with its military aircraft operating from sanctuaries in the Soviet Far East. Under Chapter VII of the United Nations Charter, the United States, when engaged in a UN peace enforcement operation, could claim the right to overfly sanctuaries used by an unannounced co-belligerent state. Whether recognized as lawful or not, however, if the aircraft was shot down, it promised an international incident of the first magnitude. (An earlier incident happened in April 1950, when Soviet fighters downed an unarmed U.S. Navy reconnaissance aircraft in international waters off the coast of Latvia, in the Baltic Sea.)[22]

The threat of a surprise nuclear strike was taken seriously. National security considerations demanded that the political risks of overflights be accepted. In late December 1950, Air Force Vice Chief of Staff General Nathan Twining briefed President Truman to that effect. After examining the overflight plans and routes, the president approved two flights, one over Russia's Arctic northern shore in eastern Siberia, and another over Russia's southern shore nearer Japan.[23] Shortly thereafter, on January 4, 1951, Headquarters USAF assigned this mission to one of SAC's newest swept-wing jet-propelled bombers, an air-refuelable B-47B then scheduled for delivery in April 1951.[24] Though initial plans called for two aircraft, only the fourth production aircraft was specially modified for reconnaissance overflights. In July it flew to Eielson AFB at Fairbanks, Alaska. While awaiting clear weather in early August 1951, this B-47B was destroyed by fire during refueling operations.[25] Later attempts to overfly the eastern USSR depended on both the production of more B-47Bs and presidential approval.

Meanwhile, Truman initiated talks on intelligence gathering with British Labor Prime Minister Clement R. Atlee and his foreign minister, Ernest Bevan (succeeded in April 1951 by Herbert S. Morrison). The American and British leaders agreed to a joint aerial reconnaissance program to overfly the western USSR during the Korean hostilities. Under terms of the agreement, apparently secured in the spring of 1951, the Royal Air Force (RAF) would reconnoiter targets in the western USSR whenever intelligence demands dictated such action. These missions would be approved by the prime minister, who turned out to be the redoubtable wartime leader Winston Churchill, reelected in October 1951. To conduct these overflights, the RAF formed a secret three-aircraft "Special Duty Flight" in July 1951. Equipped with USAF RB-45Cs, these aircraft, on two separate nighttime missions, overflew the Baltic states, Belorussia, and Ukraine: once in April 1952, and again in April 1954.[26]

With communist and UN forces on the Korean peninsula locked in a military stalemate in early 1952, U.S. leaders received intelligence that the Soviet Air Force had begun moving bombers into Siberia. These Tupolev Tu-4 aircraft, essentially carbon copies of the Boeing B-29, were flying together in large numbers into and out of airfields at Dickson on the Kara Sea, at Mys Schmidta on the Chukchi Sea, and at Provideniya on the Chukotskiy Peninsula, just across the Bering Strait from Alaska. If loaded with the nuclear weapons then believed available to the Soviet Air Force, these Tu-4s could make one-way flights to strike the United States.[27] The Office of the Secretary of Defense and the CIA again determined a need for Siberian overflights. On July 5 Headquarters USAF, which was party to these deliberations, directed SAC to modify two B-47B bombers for a special photoreconnaissance mission over "unfriendly areas."[28]

On August 12, 1952, Secretary of Defense Robert A. Lovett delivered to President Truman memoranda from General Bradley and General Walter Bedell Smith, Director of Central Intelligence (DCI), requesting two reconnaissance overflights of Soviet Siberia from Eielson AFB. After discussion, the president approved the "Northern run" between Ambarchik on the East Siberian Sea and Provideniya on the Chukotskiy Peninsula, but disapproved as too dangerous a "Southern run" over Provideniya southwestward past Anadyr to Magadan, returning eastward over the Kamchatka Peninsula. His approval of a single reconnaissance overflight, Truman told Lovett, was contingent on the concurrence of "appropriate officials of the State Department."[29] Secretary of State Dean Acheson must have concurred, because on August 15 Headquarters USAF issued instructions for the mission. It was flown successfully in October.[30] The overflight photography established that the Soviet Union was not massing Tu-4 bombers near Alaska.[31] The world that Richard Leghorn had forecast was at hand. Overflight of the USSR or other unannounced co-belligerents opposing UN peace enforcement would be approved on a case-by-case basis when the security interests of the United States demanded it.

A few weeks after the Siberian overflight, in the national elections of November 1952, Americans selected Dwight D. Eisenhower as president. In January 1953 Eisenhower took the oath of office as the thirty-fourth president of the United States. The former Supreme Commander of Allied Expeditionary Forces in Europe, who had directed the Normandy invasion during World War II and had served as the postwar SACEUR, appreciated the value of "pre-D-day photography." As a commander-in-chief intent on thwarting, if not eliminating, the threat of an atomic surprise attack, he embraced the concept and conduct of pre-hostilities strategic overflight reconnaissance. But the Korean Armistice, signed in July 1953, ended the Korean conflict and with it any legal justification for the overflights begun by his predecessor.

Eisenhower recognized the importance of strategic reconnaissance to national security, the political risks of continuing overflights for that purpose in peacetime, and the precedent set by President Truman. He decided to continue periodic reconnaissance overflights of the Sino-Soviet bloc. The revised effort, including RAF participation with the consent of his British wartime comrade Winston Churchill, comprised a major part of what was termed the SENSINT (Sensitive Intelligence) Program. (Another significant portion consisted of the reconnaissance flights around the periphery of the Sino-Soviet bloc states.) Between July 1953 and December 1956, when Eisenhower discontinued all military overflights of the USSR, the overflight component involved four USAF commands, elements of the RAF, and Republic of China (ROC) Air Force based on the island of Taiwan. In the case of the American air commands, once the president approved an overflight, authority to proceed passed from the Joint Chiefs of Staff to Headquarters USAF, then through channels to the operational unit.[32]

Meanwhile, alerted to the existence of the new Soviet Myacheslav-4 intercontinental bomber (NATO code-named BISON),[33] in March 1954 Eisenhower challenged James Killian, by now president of MIT, and other members of the Office of Defense Mobilization's Science Advisory Committee to advise him of new ways to protect America from a sneak attack. Killian formed for this task a Technological Capabilities Panel (TCP). In August, panel members began a survey of U.S. defense preparedness and of new technology that might be applied to offensive, defensive, and intelligence operations.[34] With the number of Soviet BISON bombers and nuclear weapons believed to be growing, the region of greatest concern in the USSR—from which such an aerial attack could be mounted and about which the least was known—was the Kola Peninsula in extreme northwest Russia above the Arctic Circle. Again, American and British leaders sought reliable intelligence about the type, number, and disposition of Soviet aerial forces.

On May 8, 1954, a few days after the RAF Special Duty Flight conducted its second nighttime mission, a SAC RB-47E embarked from Fairford RAFB on a daytime photographic overflight of the Kola Peninsula and various airfields east of Leningrad. Soviet MiG-17s attacked and damaged the reconnaissance aircraft during the mission.[35] Before the month's end the American-equipped RAF Special Duty Flight disbanded permanently.[36] If such strategic reconnaissance overflights were to continue with acceptable risk, another kind of airplane clearly was required, one designed expressly to fly at extremely high altitudes, far above Soviet air defenses.

TECHNICAL AND INSTITUTIONAL IMPEDIMENTS

If, in the interests of national security and after weighing attendant risks, a U.S. president authorized reconnaissance overflights of Sino-Soviet territory, what kind of aircraft was best suited to the assignment? The Strategic Air Command controlled the lion's share of Air Force reconnaissance assets because it possessed the only long-range jet-powered aircraft that could be refueled in flight and configured for deep-penetration electronic and photographic reconnaissance missions. None of these aircraft, however, was designed to operate above 45,000 feet altitude; all of them were thus susceptible to attack by air defenses. To assess the state of the art and to recommend new or improved reconnaissance technical collection systems, the Air Force turned to scientific consultants organized earlier to assist Air Force Development Planning. Among them were Carl F. J. Overhage, chief of Eastman Kodak's Color Laboratory; James Baker and Edward Purcell from Harvard University; Edwin Land, president of Polaroid; Louis Ridenour of International Telemeter; and Allen F. Donovan of Cornell Aeronautical Laboratory. At the early 1952 request of the Air Force, these men, along with Richard Leghorn, who had been recalled to active duty in the Air Force during the Korean conflict, participated in what became known as the "Beacon Hill Study," named for the Boston locale where they convened. In the years afterward, these same scientist-consultants would reappear on the Air Force Scientific Advisory Board and on other government panels.[37]

The seminal *Beacon Hill Report* appeared in June 1952. Having evaluated pre-hostilities reconnaissance requirements, the Beacon Hill group recommended improvements in sensors and identified vehicles that could fly them near or over Soviet territory. These included high-altitude balloons (then Project GOPHER),[38] high-altitude aircraft, sounding rockets, and long-range Snark or Navaho air-breathing missiles employed as drones. Although aware of the contemporary reconnaissance satellite studies at the RAND Corporation in Santa

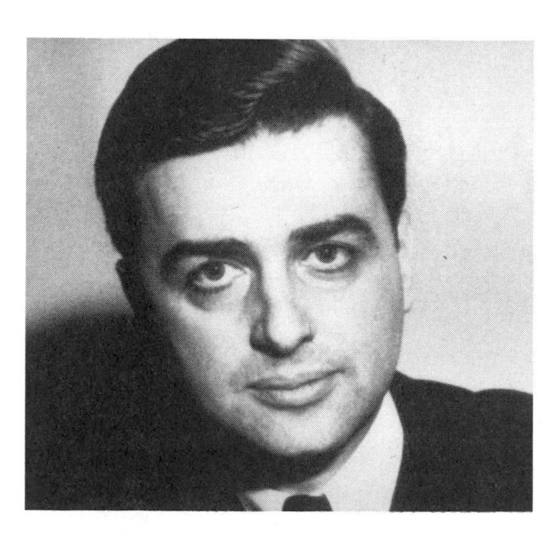

Dr. Edwin "Din" Land, president of Polaroid. Land served as a key advisor to presidents and U.S. intelligence agencies during the height of the Cold War. (Photo courtesy of the National Reconnaissance Office)

Monica, California, members of the Beacon Hill group perceived that this technology would require an enormous investment, including the development of launching rockets. Such machines could not be developed in time to meet current national intelligence demands. Both Leghorn and Land favored immediate development of balloons and aircraft that would operate at extreme altitudes of 70,000 feet or higher.[39] (The subsequent development of intercontinental ballistic missiles and the launch of an artificial earth satellite prompted both Land and Leghorn to actively support *immediate* development of reconnaissance satellites.)[40]

At the Pentagon in early January 1953, Leghorn completed his work on an Intelligence and Reconnaissance Development Planning Objective (DPO) for Colonel Bernard A. Schriever, Assistant for Air Force Development Planning. At month's end he left active duty to return to Eastman Kodak, removing himself temporarily from planning based on the pre-hostilities reconnaissance concepts he had helped formulate. The DPO called for high-altitude balloons and eventually earth satellites to provide wide-area searches of the Soviet Union, with close-area surveillance provided by high-altitude airplanes and second-generation satellites. To achieve close-area surveillance, the study identified a requirement for a special purpose, single-engine, lightweight (unarmored) reconnaissance airplane to be employed expressly for peacetime strategic reconnaissance at altitudes of 70,000 feet or higher.[41]

A few months later, on July 1, 1953, the Wright Air Development Center (WADC) issued six-month study contracts for such a special-purpose Air Force reconnaissance airplane to three East Coast firms: Bell Aircraft Company, Fairchild Engine and Airplane Corporation, and the Glenn L. Martin Company. The design requirements were the brainchild of Major John D. Seaberg. The closed procurement competition called for acquiring an airplane similar to the one Leghorn had urged the Air Force to embrace—a jet-turbine-powered aircraft that would perform "pre- and post-strike reconnaissance," possess a flight radius of 1,500 miles unrefueled, and fly at an altitude of 70,000 feet or higher. In March 1954 WADC chose for development the twin-engine Bell entry, the X-16, over the single-engine Fairchild entry. To meet near-term reconnaissance needs, the service also selected for procurement a modified version of the twin-engine British Electric Canberra bomber (later known as the RB-57) offered under license by the Martin Company. The winning X-16 design featured two wing-mounted jet engines (the second engine perceived as a safety feature), an armor-plated pressurized cabin, an ejection seat, and a single carry-through wing spar that met military specifications (MilSpecs) and required high g-loading (the ability of an aircraft to make hard, high-speed turns).[42]

Officials at Lockheed Aircraft Corporation in Burbank, California, became aware of this competition and in late November 1953 closeted themselves with their premier designer, Clarence L. (Kelly) Johnson. In March 1954, about the time ARDC selected the X-16, Lockheed sent to the recently promoted Brigadier General Bernard Schriever an unsolicited proposal for a different kind of reconnaissance airplane. Called the CL-282, it did not contain an ejection seat and featured a single jet engine, an unpressurized cabin, bolted-on, high-aspect "wet" wings, and a payload bay between them that could carry some 500 pounds in 15 cubic feet. It was designed to attain an altitude of 70,000 feet and fly for 2,000 miles at a restricted airspeed—not to exceed 390 knots (any greater speed would tear the wings off). This fragile, single-purpose, high-altitude vehicle would tolerate only very low maneuver loads—about 2.25 gs—far below the requirements of military specifications.[43]

Schriever was interested. Although the X-16 promised to meet MilSpecs, the CL-282 appeared to meet all requirements of the intelligence and reconnaissance DPO. He invited Kelly Johnson to Washington, D.C. In late March or early April 1954, Johnson and his Lockheed associates briefed Assistant Secretary of the Air Force for Research and Development Trevor Gardner, Deputy Chief of Staff for Research and Development General Donald Putt, and others. Leaders of the Air Research and Development Command (ARDC), responsible for acquisition of the counterpart X-16, successfully opposed the unsolicited Lockheed proposal, but not before it came to the attention of former members of the Air Force's Beacon Hill group now serving on James Killian's TCP. Through Philip Strong and Schriever, it also came to the attention of the CIA.[44]

Schriever knew a bright Yale-trained economist named Richard M. Bissell Jr., who recently had joined the CIA as one of Allen Dulles's principal managers. Late in May or in early June 1954, Schriever telephoned Bissell and invited him to the Pentagon. There, two members of Schriever's staff, RAND economist Burton Klein and Captain Eugene Kiefer, both of whom had continued to work on refining the intelligence and reconnaissance DPO, briefed Bissell on the CL-282 and its technical prospects and potential. Klein, working on loan to Schriever, recalled, "Bissell was immediately impressed and showed great interest in this airplane."[45]

A few weeks later the TCP began its secret assessment of the nation's offensive forces, air defenses, and intelligence capabilities. Word of Lockheed's proposed CL-282 already had reached members of the TCP's intelligence committee, including Alan Donovan and James Baker, just returned from his Air Force consultant's visit to the RAF. Chaired by Polaroid's Edwin Land and composed mostly of individuals who had served previously in the Beacon Hill Group, Land's TCP intelligence committee soon became convinced that this high-altitude airplane, believed to be nearly invisible to radar, was the nation's best answer to acquiring pre-hostilities strategic overflight reconnaissance at minimum risk.[46]

Eisenhower at that moment was ready to consider a proposal for a novel reconnaissance airplane that operated at extreme altitudes. Soviet fighters had attacked and nearly shot down the Air Force RB-47E that had flown over the Soviet Kola Peninsula a few months before. Military aircraft might again perform such provocative and illegal missions, but they increased the risk of triggering the conflict he sought to avoid. In late 1954, to provide the nation warning of surprise attack, the president decided to build and send these unarmed, single-engine Lockheed airplanes over the "denied areas" of greatest concern. At the operating level, except for Richard Bissell, Trevor Gardner, and Generals Bernard Schriever and Donald Putt, the leadership of the CIA and of the Air Force mostly opposed purchase of the CL-282. Eisenhower nonetheless determined that the CIA would direct this crucial project, with the airplane procured in the greatest secrecy outside of established channels. The Air Force would provide the needed technical support. (The project was started with Schriever's knowledge, although he learned of it later.)

On November 24, 1954, the president met at the White House to discuss the proposition with the secretaries of state and defense, the secretary of the Air Force, the DCI, and senior Air Force officers. Secretary of State John Foster Dulles "indicated that difficulties might arise out of these operations, but that 'we could live through them.'" All agreed to proceed with the project in secrecy along the lines of shared management.[47] For this joint civil-military project, the CIA would provide the funding, overall direction, and security procedures. The Air Force would provide the facilities infrastructure, trained technical personnel, and, eventually, pilots. If these decisions marked the impending demise of military aircraft overflights of the USSR, they unquestionably established strategic overflight reconnaissance as a national policy.

OVERFLIGHT AT EXTREME ALTITUDES: THE GENESIS OF CORONA

At the CIA, Allen Dulles named Richard Bissell director of Project AQUATONE, as the CL-282 effort was called. Bissell had the work under contract with Lockheed by the end of 1954.[48] The first of the famous aircraft, renamed the "U-2," was test-flown in Nevada eight months later. AQUATONE's Air Force Deputy Director, Colonel Osmond J. Ritland, supported Bissell from his office in the Pentagon, while Colonel Marion C. Mixson worked directly with Bissell at the CIA.[49] Elsewhere in Washington, in preparation for a Four-Power Summit Conference scheduled in Geneva, Switzerland, in July 1955, President Eisenhower assigned responsibility for arms control and disarmament proposals, including prospective methods for policing future agreements, to his special assistants Harold Stassen and Nelson Rockefeller.[50]

Any aerial reconnaissance overflight of another state without authorization of course remained an illegal and hostile act unless national leaders agreed to it beforehand. On July 21, 1955, at the Geneva summit conference, President Eisenhower advised Soviet leaders of just such a plan. Devised primarily by Nelson Rockefeller, assisted by Max Milliken and Walt Rostow of MIT, it became an unannounced addition to a disarmament proposal.[51] The absence of trust and the presence of "terrible weapons" among states, he asserted, provoked in the world "fears and dangers of surprise attack." To eliminate those fears, he urged that the Soviet Union and the United States provide "facilities for aerial photography to the other country" and conduct mutually supervised reconnaissance overflights.[52] Before the day ended, First Secretary of the Communist Party Nikita Khrushchev privately rejected the president's plan, known eventually as the "Open Skies" doctrine, as an obvious attempt to accumulate target information. Immediately following the Geneva conference, Harold Stassen assembled another group of experts to consider the subject further and to evaluate other arms limitation proposals.[53]

Shortly after returning to the United States, on July 29, 1955, the president publicly announced plans for launching "small unmanned, Earth circling satellites as part of the U.S. participation in the International Geophysical Year" (IGY), scheduled between July 1957 and December 1958. His statement avoided any hint at the principal underlying purpose of this enterprise, which was to establish the principle in international law of "freedom of space," with all that that implied for strategic reconnaissance conducted at altitudes above the "airspace" to which the states beneath claimed exclusive sovereignty. Crafted early in 1955 in response to TCP recommendations by Donald H. Quarles, Eisenhower's assistant secretary of defense for research and development, this initial U.S. space policy established a precedent during the IGY and would become a cardinal principle of public space law—incorporated in international treaties a decade later. It paved the way for the launch and operation of the first American reconnaissance satellites in 1959–60.[54] (See chapter 5 on the subject of freedom of space.)

In August 1955, at the invitation of World War II hero and presidential confidant James Doolittle, Richard Leghorn became a member of the Stassen/Rockefeller arms control and disarmament group, serving on the "aerial inspection" subcommittee. Now aware of the U-2 project, Leghorn viewed strategic reconnaissance as a potential "inspection system" that would serve two critical functions: to forewarn of surprise attack and to supervise and verify arms-reduction and nuclear test-ban agreements.[55] President Eisenhower and other key administration officials also embraced this view before the first U-2 mission ventured into Soviet airspace.

In the meantime, after a prolonged gestation period, the Air Force completed testing of its balloon reconnaissance system. This effort mounted aerial cameras in gondolas suspended beneath large polyethylene balloons for flight at very high altitudes (greater than 70,000 feet). The balloons, launched from Western Europe, were intended to drift over the USSR on prevailing winds and to be recovered in Japan or Alaska. This effort perfected methods of flying balloons at constant-pressure altitudes, parachutes capable of supporting a 600-pound load descending from high altitudes, and an aerial recovery system that employed C-119 cargo airplanes equipped with grappling lines and winches able to snatch the gondola packages in mid-air.[56] The camera employed in this project was designed and engineered for the Air Force by Walter Levison and Francis Madden at Boston University's Optical Research Laboratory. Weight constraints precluded a standard trimetrogon installation, and the two men fashioned a novel lightweight camera that used a 9 × 9-inch film format, equipped with two separate 6-inch Metrogon lenses that viewed the earth obliquely. The lenses provided a 10-degree overlap at the center and the camera covered the ground from horizon to horizon. Various firms manufactured some 2,500 of these BU "duplex cameras" for the project.[57]

This initial balloon reconnaissance project, known as GENETRIX (WS-119L), was executed in early 1956. Teams from the Strategic Air Command launched the balloons from bases in West Germany, Scotland, Norway, and Turkey beginning on January 10. Between that date and February 6, when President Eisenhower terminated the effort in the face of strong Soviet protests, 516 balloons were released to sail over Russia on the prevailing winds. The U-2 aircraft was about to begin flight operations, however, and, not wanting to alert Soviet leaders to its high-altitude capabilities, administration officials directed the Air Force to ballast the GENETRIX balloons to prevent them from ascending to altitudes greater than 50,000 feet. As a

Walter Levison, who developed the WS-461L balloon camera and proposed that the camera be modified for use in a reconnaissance satellite. Levison's camera design was used on the CORONA satellite. (Photo courtesy Walter Levison)

result, numerous balloons with their camera packages were shot down; only sixty-seven reached the recovery area, and of these only forty-four were retrieved.[58]

A few months after Eisenhower terminated that phase of GENETRIX, he approved the first flight of Project AQUATONE. On July 4, 1956, a camera-equipped U-2 took off from Wiesbaden, West Germany, to survey the USSR's naval shipyards and especially its submarine construction program. It overflew Poland, Belorussia, Moscow, Leningrad, and the Soviet Baltic states. Contrary to American expectations, Soviet radar detected and tracked this first U-2 at its 70,000 feet altitude.[59] This overflight caused considerable consternation among the post-Stalin Kremlin leaders. Strategic reconnaissance, to be sure, furnished not only indications and warning but also targeting data for a nuclear attack. According to his son, Sergei, Soviet Communist Party Chairman Nikita Khrushchev rejected the 1955 Open Skies proposal because he believed Americans were

> really looking for targets for a war against the USSR. When they understand that we are defenseless against an aerial attack, it will push the Americans to begin the war earlier . . . [and] if in this fear of each other the Americans realized that the Soviet Union would become stronger and stronger, but was weak now, this [intelligence] might push them into a preventive war.

The event triggered Kremlin orders for new surface-to-air missiles and high-performance fighters, and accelerated work to perfect an intercontinental ballistic missile.[60]

U.S. leaders in 1956 might have been seeking the U-2 to collect intelligence of Soviet military capabilities that would warn them of an impending nuclear Pearl Harbor. But the flights forcefully reminded Khrushchev and other Kremlin officials of the Luftwaffe reconnaissance overflights that preceded Germany's surprise attack on the USSR in 1941. Fears of a surprise attack among Soviet authorities were essentially mirror images of those shared by their American counterparts.

In the months that followed, President Eisenhower approved only a small number of U-2 overflights. These assayed Soviet nuclear production and test facilities and the number, kind, and disposition of its military forces. The intelligence thus acquired might also dispel or confirm American intelligence concerns that "bomber" and "missile" gaps existed between the two superpowers. Soviet scientists and engineers, in the meantime, successfully tested an ICBM in August 1957 and, two months later, launched the world's first artificial earth satellite, Sputnik I. In addition to its profound psychological impact among the publics of Western nations, Sputnik's orbiting of the earth did indeed establish the international precedent of "freedom of space" and of the eventual right of overflight in outer space that the president and his advisors had desired. But the media clamored for space projects with which to surpass the Russians. Without divulging the intent of his IGY-generated space project and its clandestine policy, the president had to decide who would control and direct U.S. astronautical activities, now certain to be much larger and more diverse than anyone had imagined.

Working with his advisors and the Congress, Eisenhower answered these questions between late 1957 and 1961. The answers turned primarily on issues of national security.[61] The issue of first concern arose a few weeks after the October 4 launch of Sputnik I. On October 24 the President's Board of Consultants on Foreign Intelligence Activities (PBCFIA) submitted one of its periodic reports to Eisenhower. Formed by executive order the year before to review and report to the president on activities of the government's intelligence organizations,[62] this eight-member board was chaired by the ubiquitous James Killian and included among its members Edwin Land. The PBCFIA report recommended an evaluation of overhead reconnaissance systems, including satellites.[63]

The U.S. Navy and Army Air Force had begun studies of earth satellites in 1945–46. The RAND Corporation had continued them for the U.S. Air Force after 1947. On March 1, 1954, RAND issued a concluding Project Feed Back report which recommended that the Air Force begin a reconnaissance satellite project. Simultaneously, the Defense Department Study Group on Guided Missiles advised the Air Force that the design and construction of an ICBM was technically feasible, and that if such a program was pressed vigorously, ICBMs would become available by 1960. The group recommended that the program begin immediately. On February 16, 1954, the Air Force transmitted these findings and recommendations to Assistant Secretary of Defense for Research and Development Donald Quarles. Endorsed by the Eisenhower administration, by year's end Bernard Schriever found himself in charge of the Western Development Division (WDD), the ARDC organization created specifically to build the nation's ICBM.[64] Starting on parallel tracks, the military satellite and its potential booster would soon converge on the West Coast—all under Schriever's direction.

At the ARDC's Detachment 1 at Wright-Patterson Air Force Base (WPAFB), the RAND Feed Back report particularly impressed Major Quenten A. Riepe who, along with Captain James S. Coolbaugh, promoted the project at ARDC and at Headquarters USAF. The Air Force's Geophysics Laboratory and the Rome Air Development Center in New York supported this effort. In November 1954, ARDC issued System Requirement Number 5, which called for competitive system design studies of a reconnaissance satellite. On March 16,

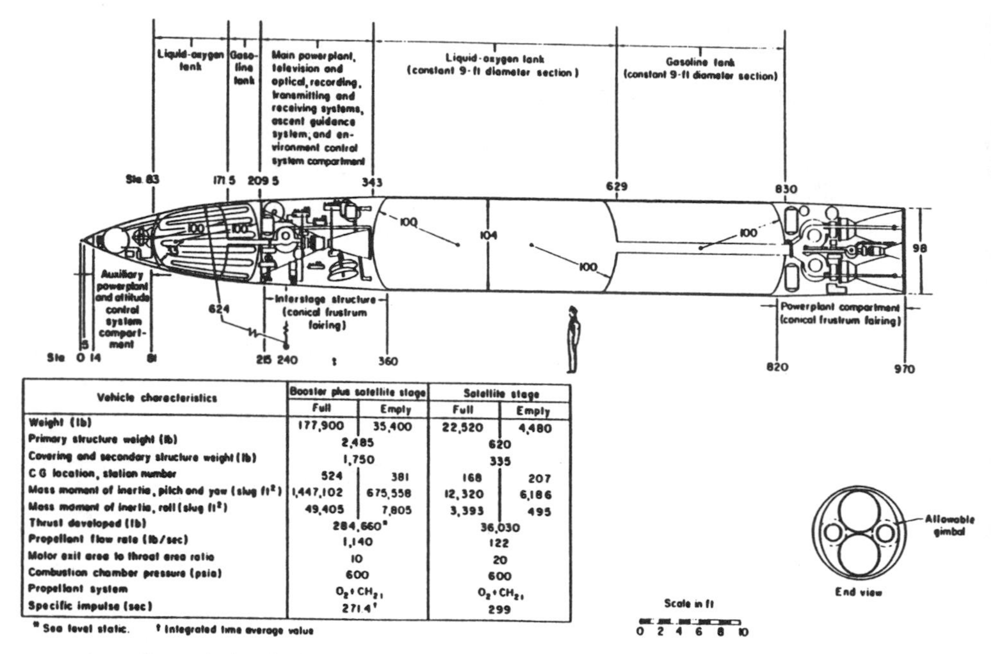

Vehicle characteristics	Booster plus satellite stage		Satellite stage	
	Full	Empty	Full	Empty
Weight (lb)	177,900	35,400	22,520	4,480
Primary structure weight (lb)	2,485		620	
Covering and secondary structure weight (lb)	1,750		335	
C G location, station number	524	381	168	207
Mass moment of inertia, pitch and yaw (slug ft²)	1,447,102	675,558	12,320	6,186
Mass moment of inertia, roll (slug ft²)	49,405	7,805	3,393	495
Thrust developed (lb)	284,660*		36,030	
Propellant flow rate (lb/sec)	1,140		122	
Motor exit area to throat area ratio	10		20	
Combustion chamber pressure (psia)	600		600	
Propellant system	$O_2 + CH_{2.1}$		$O_2 + CH_{2.1}$	
Specific impulse (sec)	271.4†		299	

* Sea level static. † Integrated time average value

A satellite vehicle schematic diagram from the RAND "Project Feed Back" report, March 1954. RAND engineers envisioned a nuclear-powered satellite with a television camera. (Courtesy of RAND)

1955, Headquarters USAF issued General Operational Requirement (GOR) Number 80 (SA-2c). GOR 80 approved construction of, and provided technical requirements for, a reconnaissance satellite.[65]

At ARDC's Detachment 1 in the spring of 1955, Riepe directed a four-member team that worked on the reconnaissance satellite project. In addition to Riepe and Coolbaugh (now technical director, but specializing in auxiliary power and propulsion), it consisted of First Lieutenant John C. Herther (guidance and stabilization) and Captain William O. Troetschel (communications, command and control). Lieutenant Colonel William G. King replaced Riepe as the satellite project director in mid-August and, shortly thereafter, the Detachment 1 satellite team awarded small contracts to various firms for improvements in guidance systems and auxiliary power. Funding was obtained with which to pay four firms for a one-year competitive design study of a reconnaissance satellite, and three responded positively: Radio Corporation of America, Glenn L. Martin Company, and the Lockheed Aircraft Company. On the West Coast, made aware that the reconnaissance satellite project would affect directly the procurement of ICBMs, Schriever arranged in October to transfer the Air Force reconnaisance satellite office from WPAFB to his own WDD in Los Angeles.[66]

The original WS-117L team on March 4, 1956, shortly after their arrival at the Western Development Division, Los Angeles. Standing (left to right): Capt. William O. Troetschel; Edwin Kolb; 1st Lt. John C. Herther; Lt. Col. William G. King; Russell Johnson; James Suttie; Joseph Fallik; Capt. James S. Coolbaugh; and Capt. Frank S. Jasen. Kneeling: Fritz Runge; Capt. Richard P. Berry; Navy Capt. Robert C. Truax; Robert Copeland; Lt. Col. George P. Jones; Lt. Col. George Harlan.

Except for Lieutenant Colonel King, who was placed in charge of the Snark Project and remained at WPAFB, the satellite team moved to Los Angeles in February 1956. King, however, ensured that his team would remain together, be assigned to WDD, and not report to Ramo-Wooldridge, the firm that oversaw ballistic missile developments for Schriever. At the urging of Trevor Gardner, Undersecretary of the Air Force for Research and Development, Schriever named Commander Robert C. Truax, USN, as director of the Air Force Reconnaissance Satellite Office. Air Force Colonel Frederic C. E. (Fritz) Oder succeeded Truax in August, but Truax stayed on as his deputy. The Lockheed Aircraft Corporation won the satellite design competition in June and received a letter contract in October 1956. It would develop space reconnaissance and missile early warning satellites for the Air Force in what was now called the Weapons System (WS) 117L Program. By 1957, work on WS-117L was underway at Lockheed's newly formed Missiles and Space Division, then located in Palo Alto, California.[67]

The Air Force WS-117L program represented the nation's only reconnaissance satellite effort in 1957; it continued to labor under strict funding restrictions imposed by Secretary of the Air Force Donald Quarles.[68] Planning for space reconnaissance featured a second stage booster-satellite to be launched into a near polar orbit by an Atlas ICBM adapted for that purpose. At first designed for a long-life mission of one year at an atmospheric drag-free altitude of 300 miles, the WS-117L Lockheed "Agena" satellite, as it later became known, was based on the 1954 RAND Feed Back study and was originally conceived to be gravity gradient–stabilized. That is, the vertically oriented Agena moved on orbit with its fixed, nose-mounted Eastman Kodak strip camera pointed toward the earth, thus aligning the long axis of the satellite's mass distribution perpendicular to the earth. (The WS-117L "pioneer" version of the Eastman camera was designed to deliver a resolution of 100 feet at the earth's surface; a more advanced version promised a resolution of 20 feet.) The gravity gradient stabilization scheme eliminated the need for fuel and associated weight required for gas jets, and made possible long-lived, very high-altitude orbital operations. It required only electricity for momentum-wheel damping with rate-sensing gyroscopes. Power was to be supplied by batteries recharged through solar cells. Film would move across the camera slit in the opposite direction of vehicle motion. The exposed film was to be processed on board, scanned electronically with a CBS "flying-spot scanner," and the video signal transmitted to Earth when the satellite passed within sight of a ground station in the United States. There, the signal could be reformed into a photographic image of the original scene.

This approach to space reconnaissance, in the view of some at the Air Force–funded RAND Corporation in Santa Monica, had significant drawbacks. First, the satellite had to carry enough film to be used over a one-year lifetime. Second, without a recorder for storing the images, the film had to be read out on each pass and then discarded on orbit. Third, the limited radio bandwidth and data transmission rate, coupled with the brief time available for communication while the satellite remained in line of sight of a ground station as it passed overhead, seriously restricted the number of images that could be relayed to Earth. In fact, RAND calculations yielded a daily figure equivalent to five or six 9 × 9-inch photographs whose quality was 100 lines per millimeter transmitted to Earth.[69]

Back in 1956, Richard C. Raymond in RAND's electronics division had compared the long-lived film readout system with a recoverable payload in which exposed film would be returned to Earth in a reentry capsule after a short mission. "Film recovery," Raymond calculated, "would yield at least two orders of magnitude more data" in a given period of time. RAND sent these findings to Headquarters USAF in March 1956. The RAND proposal, however, also had a drawback of its own. When seeking indications and warning of a potential surprise attack against the United States, one wanted information in near real time, not days later in the form of exposed reels of film that required developing. Rejecting the RAND proposal, the Air Force contracted with Lockheed for the WS-117L visual readout, infrared, and ferret (signals intelligence) system a few months later, in October.[70]

Amrom H. Katz, the Air Force physicist who had participated in the 1946 CROSSROADS atomic bomb tests and had joined RAND several years later, was a recognized camera expert and a champion of pursuing simple technical solutions rather than complex ones. Katz and a RAND associate who shared his views, Merton E. Davies, embraced Raymond's answer to a fast-paced space reconnaissance project. While attending the annual meeting of the American Society of Photogrammetry in Washington on March 4–6, 1957, Davies encountered Frederic Wilcox of Fairchild Camera and Instrument Corporation. Wilcox described for him a new Fairchild panoramic camera developed for an aerial drone. It fit inside a pod and the entire camera rotated in a drum. Davies thought about the complexity inherent in this design. By the time he and Katz boarded a plane for the flight back to the West Coast, he had a "hot idea" for space flight: fix the camera to the satellite and spin the entire ensemble.[71] By late spring 1957, Davies and Katz were advocating the spinning camera and film-recovery scheme in briefings for scientific and military officials who visited RAND. These included Colonel Fritz Oder, Air Force director of WS-117L, his deputy, Navy Commander Bob Truax, and eventually members of the Reconnaissance Panel of the Air Force Scientific Advisory Board (SAB), the SAB's ad hoc Panel on Advanced Weapons Technology and Environment, the Defense Department's Advisory Group on Special Capabilities (Stewart Committee), and those of the Science Advisory Committee of the Office of Defense Mobilization. James Killian and Edwin Land were among the members of the last group. The two RAND champions of a recoverable space reconnaissance capsule system completed their formal study, with the assistance of other RAND coauthors, and issued it on November 12, 1957.[72]

Though aware of both the planned film readout and recovery space reconnaissance systems by mid-1957, Killian and Land still favored ongoing aerial overflight reconnaissance to meet the nation's immediate intelligence requirements. Both had recognized reconnaissance satellites as a system of future consequence in the 1955 TCP report. They had recommended that the nation prepare for this activity by establishing in public international law the principle of "freedom of space." To that end, the president had approved the IGY satellite project and its related space policy.[73] The launch of Sputnik I on October 4, 1957, which helped establish the legal precedent they sought, also caused them both to reconsider the timing and technical prospects of space reconnaissance. On October 24, reporting as members of the President's Board of Consultants on Foreign Intelligence Activities, but after hearing the satellite briefings by RAND and WS-117L officials, they advised Eisenhower that neither a new reconnaissance aircraft under study at the CIA (Project OXCART, which would become the supersonic A-12/SR-71) nor the Air Force WS-117L readout reconnaissance satellite would achieve operational status before 1960. They recommended evaluating the interim solution proposed for an advanced reconnaissance system: the film-recovery satellite advocated at RAND and adopted by the WS-117L office.[74]

By the fall of 1957, both Killian and Land had Eisenhower's complete confidence. Project AQUATONE's U-2 had proved a stunning technical success and an intelligence bonanza. If a satellite that ejected a film capsule for recovery on Earth appeared promising to them, it would be investigated. On October 28 "the Executive Secretary of the National Security Council notified the Secretary of Defense and Director of Central Intelligence that the president had asked for a joint report from them on the status of the advanced [reconnaissance] systems." On December 5 Undersecretary of Defense Donald Quarles, another trusted presidential confidant, replied on behalf of the Defense Department and of DCI Allen Dulles. Because of the sensitivity of the subject, he said, his review would be conducted through oral briefings. Held later in the month, the technical evaluation apparently confirmed the views of Killian and Land. The WS-117L satellite that took pictures, developed them on board, then scanned the film electronically and radioed the images to stations on Earth faced daunting, long-term technical challenges. *If* atmospheric reentry techniques based on ICBM warhead technology could be perfected, film exposed on one- or two-day missions might be returned to Earth for developing and analysis. The parachutes and air-sea recovery system already perfected to retrieve GENETRIX cameras and film added confidence in this approach.[75]

The interim reconnaissance satellite system that Quarles compared with the WS-117L readout system was the same plan that had been described in November by Davies and Katz, and had already been endorsed by Fritz Oder and Bernard Schriever. It consisted of a Thor IRBM liquid-propellant launch vehicle with an Aerobee 75 solid-propellant second stage developed for the Vanguard IGY satellite launcher. Mounted atop the Aerobee was a football-shaped third-stage satellite that contained the camera, film, and a small solid-propellant recovery rocket. The spin-stabilized satellite contained a Fairchild-manufactured transverse panoramic slit camera that featured a 12-inch focal-length f/3.5 lens, and could cover a narrow angle of approximately 21 degrees. Wide-angle scanning, accomplished by spinning the satellite, moved the lens across the field during the exposure time. Pictures were to be taken only when the lens, mounted perpendicular to the roll axis, swung past the earth below. At 135–140 miles altitude, the camera would produce a resolution on the surface of 60 feet at 40 lines per millimeter.[76]

On the West Coast, Oder had counted himself among the first Air Force converts to the Davies and Katz recoverable reconnaissance satellite concept back in the summer of 1957. Indeed, by August he had sold the concept to his superiors, new Major General Bernard Schriever, commander of the Air Force Ballistic Missile Division, and Schriever's deputy, Brigadier General Osmond Ritland. The three men also conferred with Richard Leghorn, by now a member of the Aerial Inspection Subcommittee of the President's Arms Control and Disarmament Group. All agreed. If this effort was to succeed quickly, it would require presidential approval and the highest of national priorities, and needed to be prosecuted covertly like Project AQUATONE. Schriever would soon approach select members of the Air Staff and others in Washington about this project and about a "second story" that might be devised to provide a cover. Oder's WS-117L program office in Inglewood, California, meanwhile, included the Thor-boosted reconnaissance satellite, identified as Program IIA, in its 1957 WS-117L Development Plan. Lockheed also was instructed to plan for this addition to the WS-117L program.[77]

In Washington, D.C., on November 12, 1957, Assistant Secretary of the Air Force for Research and Development Richard E. Horner carried out his own technical assessment for Secretary of Defense Neil McElroy. He affirmed that a film-recovery satellite could reach operational status at least a year before the WS-117L readout satellite.[78] A few days later, on November 15, in the aftermath of the second, more spectacular Soviet Sputnik launching, Eisenhower named James Killian to be his special assistant for science and technology and to serve as chairman of the new President's Science Advisory Committee (PSAC). In his new capacity as presidential science advisor, and still serving as chairman of the PBCFIA, in early December Killian conferred at the White House with Polaroid's Land, the CIA's Bissell, President Eisenhower's staff assistant Army Colonel Andrew Goodpaster, and Schriever. The men reviewed aircraft strategic reconnaissance capabilities and the potential options for satellite reconnaissance. A film-recovery satellite acquired and managed through a covert program, they concluded, represented the nation's best near-term choice to augment the U-2. A Thor IRBM and the Lockheed liquid-propellant Agena booster-satellite developed for the WS-117L, they agreed, enabled a heavier payload. The larger and more powerful Lockheed upper stage, which could be stabilized on all three axes in space, would be substituted in place of the smaller solid-propellant Aerobee 75.[79]

On February 6, 1958, a few weeks after Quarles's December review of strategic reconnaissance systems, Killian and Land met with DCI Allen Dulles, Secretary of Defense McElroy, and Undersecretary of Defense Quarles. They agreed to separate the Air Force film-recovery satellite Program IIA from the WS-117L program and assign it to a CIA–Air Force team, again led by Bissell. The next day, on February 7, Killian and Land met with Eisenhower to discuss the plan. Land explained for the president that they could expect a lower resolution of objects at the earth's surface in photographs taken from space, compared with the resolution obtained in photographs taken by high-altitude balloons and aircraft. But the proposed film-recovery satellite, he said, did not radiate any electronic signals and would be almost undetectable. The Air Force WS-117L readout reconnaissance satellite program, which already had received a good deal of publicity, would continue, thus providing the Air Force and its WS-117L contractors an opportunity to surmount the technical challenges and deliver near real-time images.

After listening to the recommendation, the president agreed that an interim film-recovery satellite project should begin, but independently and covertly, separated from the larger reconnaissance satellite program and managed in a manner like the U-2. The CIA, Eisenhower emphasized, should be in charge and the new Advanced Research Projects Agency (ARPA) should execute its orders.[80] In February 1958, with scant experience in or technical skills associated with launch vehicles and automated satellites, the CIA was thus charged with managing a crucial space project. The president's decision represented his preference for civilian control of national intelligence and his confidence in the men who had quickly and successfully discharged Project AQUATONE. That preference and confidence notwithstanding, to execute the space reconnaissance project the agency unquestionably would have to depend on General Schriever and the team he had assembled at the Air Force Ballistic Missile Division and its aerospace contractors.

FIRST YOU SEE IT AND THEN YOU DON'T: CORONA UNDERWAY

A new entrant in the civil-military space arena, the Advanced Research Projects Agency (ARPA) was established in the Defense Department on February 7, 1958. The president assigned to ARPA, along with military space activity, temporary responsibility for all U.S. civil satellite projects. With the authority and responsibility for directing astronautical ventures thus consolidated, Eisenhower hoped this new agency might eliminate the interservice feuding over seemingly glamorous space missions. In military space matters, the Air Force now had to respond to ARPA orders in developing and conducting space flight operations. Temporarily, ARPA also would be involved in the covert satellite reconnaissance effort. It would openly fund Air Force procurement of the Thor boosters and Agena upper-stage satellites. The CIA would provide the security system and covertly procure reconnaissance components, such as the cameras and film-reentry capsules. The Air Force would furnish the overall management and technical infrastructure.[81]

Bespeaking the enormous influence that Eisenhower's scientist-consultants now exerted in the administration, a day or two after meeting with the president, Polaroid's peripatetic Edwin Land visited CIA headquarters and informed a startled Richard Bissell that he would now direct a covert reconnaissance satellite project. DCI Allen Dulles, to be sure, knew of his subordinate's impending assignment, but it was Land who told Bissell of his new responsibility. At the CIA, in addition to the covert Project AQUATONE for which he also served as director, Bissell held the official title of Special Assistant to the DCI for Planning and Development. Bissell drew the CIA cadre for the satellite project from his Development Projects Staff. Before month's end he confirmed as his deputy director Air Force Brigadier General Osmond J. Ritland, Schriever's vice commander at the Air Force Ballistic Missile Division, who had served so ably as his first deputy on Project AQUATONE. Under Ritland, the Air Force once again would furnish the project infrastructure, in this case developing, launching, commanding, and controlling all of the satellites on orbit in addition to providing recovery of the film capsule above and, in conjunction with the U.S. Navy, on the surface of the Pacific Ocean.[82]

Before work on the project could proceed, Eisenhower officials first had to eliminate the publicly known WS-117L Thor-based reconnaissance satellite film-recovery Program IIA. Next, they had to resurrect it as a covert satellite project with a plausible "cover" (or "second story") to account for its existence. Finally, Richard Bissell had to assemble and organize the contractor team that would execute the covert satellite project. In the first instance, Herbert York, ARPA's chief scientist, followed the instructions of Undersecretary of Defense Donald Quarles. Quarles prepared a directive that was signed by ARPA's newly named director, Roy W. Johnson, and sent to Air Force Secretary James H. Douglas Jr. on February 28, 1958. It canceled the Air Force Thor-boosted reconnaissance satellite recovery component of the WS-117L program and authorized in its place the Air Force Discoverer Project, which would develop a "biomedical capsule" for the recovery of biological specimens lofted into space atop Thor-Agena launch vehicles. This new scientific biomedical space project, the directive asserted, was expected to contribute to America's early achievement of manned space flight.[83] It was Quarles who "set this all up," York recalled, and who pulled the strings of this public sleight-of-hand.[84]

Simultaneously, at the monthly review meeting of Air Force and contractor participants, John H. (Jack) Carter, Lockheed's manager of WS-117L, announced without explanation that Program IIA had been canceled. As one attendee recalled, RAND's Amrom Katz and Merton Davies, the two men who had fashioned that program, were in the audience and they jumped out of their chairs:

> They went ballistic, Amrom particularly. Amrom took it upon himself to try to get the effort reinstated and he began going around the country briefing anyone who would listen about the unwise decision to cancel the recoverable camera system. I mean he had a cause! He became so well known as an agitator on this that he disqualified himself for being cleared for what was now a black program—even though he had conceived it! The folks in charge knew that if they cleared Amrom, he would immediately cease agitating and that would tip everyone else that the program was underway.[85]

During the next six weeks Bissell identified the contractor team and its leadership and selected a name for the effort: Project CORONA. The name came first, confirmed at a meeting of project officials on March 10, 1958.[86] On March 15 Bissell met with General Ritland and confirmed the choice of the Douglas Aircraft Thor booster and Lockheed Agena second stage for CORONA. Moreover, they determined to make Lockheed the system engineer, responsible for the technical direction and integration of the entire effort. Lockheed's Project CORONA manager therefore would have to be one of its best engineers. Finally, they discussed funding an alternate camera to backstop the primary Fairchild–General Electric system.

The decision to consider a backup camera arose at least in part in response to an unsolicited proposal that Bissell had received a month before from a new firm: Itek (Information Technology). The firm's founders, Richard Leghorn as president and Duncan Macdonald and Arthur Tyler as vice presidents, had incorporated on September 27, 1957. Tyler and Leghorn were Eastman Kodak executives, and Macdonald now was Boston University's Dean of the Graduate School (Tyler had invented the Eastman Kodak "Minicard" system, an IBM punched card that contained a high-resolution microfilm negative, which might be a photographic image or an engineering drawing, but could be sorted and retrieved by a computer.) On October 4 Sputnik I shocked the world. For the nascent "document retrieval" company, that shock provided the impetus for a rapid takeoff. At the financial closing on October 10, Itek's founders put up only a modest amount of money while Laurance S. Rockefeller and other Rockefeller family members furnished most of the funds required for the first few months of operation. Within weeks, Leghorn negotiated a subcontract with Ramo-Wooldridge to develop and manufacture equipment that would process and catalogue the images produced by the WS-117L readout reconnaissance satellites. The firm now began to hire personnel for this effort.[87]

In the meantime, the Air Force had advised Boston University that it would cease funding operation of the BU Physical Research Laboratories (BUPRL—formerly the BU Optical Research Laboratory) that since 1946 had operated under contract to Brigadier General George Goddard's Aeronautical Photographic Laboratory at WPAFB. Directed by F. Dow Smith, chairman of the BU physics department, with Walter Levison serving as his deputy, the nonprofit BUPRL over the years had designed and developed numerous aerial cameras for the Air Force.

Indeed, in early 1957, Levison had begun the design of a panoramic camera for a follow-on Air Force balloon reconnaissance project called WS-461L. Using the higher resolution of the center lens field to sweep a wide angle, the panoramic camera provided high resolution over a wide swath. Such cameras, to be sure, had been used since the nineteenth century, but except for James Baker's work on the spherical shell lens, little effort had been invested in developing aerial panoramic cameras. In 1949, Lieutenant Colonel Richard W. Philbrick, the Air Force Liaison officer at BUORL, for the first time modified an S-7 Sonne strip camera and mounted it so that the film traveled perpendicular to the direction of flight. He rotated the entire camera around the longitudinal axis of the aircraft and pulled the film past a slit synchronously with the rotation of the camera body and lens. A dramatic panoramic photograph of Manhattan Island taken with this camera appeared in an issue of *Life* magazine that year. Philbrick's pioneering work had prompted the Fairchild design of a rotating aerial camera that caught the attention of Davies and Katz; however, Levison avoided rotating the camera by selecting a 12-inch focal length f/5 triplet, three-element lens and rotating it 120 degrees about its rear node, back and forth, perpendicular to the line of flight. Again working with Francis Madden and using this mechanization and the fast (for aerial photography) f/5 lens, this camera, when combined with high-resolution 70mm (2-inch) document copy-type film, would immediately produce images with 100 line pairs per millimeter. In March 1957 Katz christened the proposed camera design HYAC (for high acuity), and, on viewing the test results, in January 1958 he declared: "Seeing this photo . . . gave me a real and honest thrill. I think this is one of the top achievements in the history of aerial photography, certainly that portion of the history to which I've been exposed and in which I've been involved."[88]

The HYAC camera underwent flight tests in late 1957. At the same time, Itek's president Richard Leghorn met with BU's president Harold C. Case. Although the USAF might still fund specific projects like the HYAC camera, Case feared that the overall expense of operating the BUPRL would quickly drain the university's financial reserves. He wanted to divest the university of the enterprise immediately. Moving more rapidly than other bidders, again with the financial backing of Laurance S. Rockefeller, Leghorn made an offer that acquired for Itek the entire 106-member staff of the BUPRL and all of its on-going contracts, physical equipment, camera designs, and research reports, effective January 1, 1958.[89] At its core, Itek now was BUPRL resurrected to operate for profit.

Leghorn and Macdonald were aware of the impending Air Force–CIA satellite reconnaissance project and its planned use of the Fairchild panoramic camera. With the flight-tested HYAC camera behind them, they possessed an impressive alternate. If scaled-up from a 12-inch to a 24-inch focal-length lens, and using high-resolution 70mm film, calculations showed that this nodal point-scanning, 70-degree panoramic camera would provide a resolution on the earth's surface of 20 feet. That was a significant improvement over the 60-foot resolution of the Fairchild spin-stabilized camera. Moreover, with sufficiently low blur rates, faster optics, and projected Eastman Kodak film improvements, a scaled-up HYAC-type camera might achieve a resolution at the earth's surface approaching that of balloon-borne cameras. The Itek camera proposal, which arrived at CIA headquarters in mid-February 1958, prompted Bissell and Ritland to consider funding a backup to the Fairchild camera. But were other American aerial camera systems available that might be preferred in the space reconnaissance role?

To answer that question, on March 18, 1958, CORONA leaders conducted an evaluation of alternate cameras at the Old Executive Office Building in Washington, D.C.[90] In addition to Bissell and Ritland, the assessment panel consisted of the president's science advisor, James Killian, and two of his key PSAC advisors, Edwin Land and Harvard chemist George Kistiakowsky. The remaining panel members included Herbert York, ARPA's chief scientist, and the Air Force WS-117L managers, Fritz Oder and his deputy, Bob Truax, who had just moved to Washington for an ARPA assignment that would in fact cover his new role as a technical advisor to Richard Bissell on Project CORONA.

Four companies offered an alternate camera at this review: General Electric, Fairchild, Eastman Kodak, and Itek. The presenters for each of the firms arrived separately and waited in different anterooms, and each of the teams briefed the assembled CORONA evaluators alone. General Electric had hired Richard Raymond from RAND, and that firm offered a variation of the Fairchild spinner. Fairchild, in turn, offered a refined version of its original camera that could, it was hoped, secure a resolution at the earth's surface somewhat better than the 60 feet claimed for the original. Eastman Kodak, which held the contract for the pioneer and advanced strip cameras of the WS-117L readout system, likewise recommended a version modified for panoramic coverage with spin stabilization. Finally, physicist and Itek cofounder Duncan Macdonald and John C. (Jack) Herther offered a reciprocating 70-degree field panoramic camera with an f/5 Tessar-type 24-inch focal length lens, otherwise similar to the high-performance HYAC balloon camera.[91]

Itek's proposed vertical-looking camera scanned at right angles to the line of flight, which demanded a satellite horizontally stabilized on all three axes. That introduced technical complexity and accounted for the presence of Jack Herther. Richard Leghorn had hired him as one of Itek's first technical employees just before the BUPRL acquisition. A 1955 MIT graduate, he wrote his thesis on a gyrostabilized ascent guidance system for the WS-117L orbiting stage, after which he had been posted as a reserve officer to the nascent program office in the ARDC's Detachment 1 at WPAFB.[92] Now Herther explained for the Project CORONA evaluation team how the Lockheed ascent guidance system could be modified to stabilize the Agena horizontally on all three axes in space for a short duration, low-altitude reconnaissance mission. This orbital attitude-control system, Herther affirmed, would produce the pointing accuracy and low roll and pitch blur rates needed for the Itek camera to deliver a resolution at the earth's surface of at least 20 feet.[93]

Duncan Macdonald had worked previously with Arthur Lundahl, chief of the CIA's photo-interpretation unit, on matters of high-altitude Air Force balloon and U-2 aerial photography. He knew that a camera's effective resolution at the earth's surface allowed photo-interpreters to positively identify objects 3 to 5 times larger than the resolution achieved.[94] Based on photo interpretation needs, the performance experience with the HYAC flight test program, and a stable (low blur rate) platform in space, Macdonald predicted that eventually it should be possible to achieve balloon quality photographs from satellite altitudes. After all of the presentations, the CORONA evaluation team conferred and selected Itek. The long-shot newcomer would supply the backup camera.[95]

On March 24–26, 1958, Bissell and Ritland closeted themselves with all of the primary CORONA contractor representatives at the Flamingo Motel in San Mateo, California. Bissell informed attendees that a backup camera would be procured from Itek. Lockheed announced that James W. Plummer, formerly in charge of the WS-117L Eastman Kodak payloads, would serve as the CORONA manager responsible for the technical integration of the project. Project participants agreed that General Electric would provide the recovery system and that the effort would consist of ten CORONA vehicles, with three more if needed, launched from Vandenberg AFB on the California coast. Component fabrication, assembly, testing, and a first launch, participants agreed in a burst of optimism, could be accomplished before the end of 1958.

Back in Washington, D.C., on April 9, 1958, Bissell finished for the president's approval the CORONA Project Proposal. It called for the concurrent procurement of both the Fairchild and Itek cameras, though at this point the Itek system appeared a clear favorite because of its better initial resolution and promise of even greater resolution for photo-interpretation growth potential. Two days later, perhaps at the urging of Edwin Land, General Ritland and Bissell decided against procuring the Fairchild camera with its spin-stabilization, and in favor of the Itek HYAC-type camera that required a stable platform in space. The revised proposal outlined a project that would consist of twelve launchings, become operational in June 1959, and conclude a year later in June 1960 when the WS-117L readout system was scheduled to become operational.

Fairchild would remain in the project, at least temporarily, by fabricating the Itek-designed cameras.[96]

The revised CORONA Project Proposal also identified ARPA as exercising overall technical supervision, with the Air Force, through Air Force Ballistic Missile Division, acting as its agent. The CIA would remain responsible for CORONA's security system and for procuring the reconnaissance equipment. With the concurrence of ARPA director Roy Johnson and other project participants, Bissell and the Deputy Director of Central Intelligence, General Charles P. Cabell, presented this proposal to President Eisenhower on April 16, 1958.[97] After asking some questions, the president verbally approved it.[98] On April 25 Bissell issued a two-page Statement of Work to guide the prime contractor, Lockheed's Missile and Space Division. Among other objectives, it called for photographs with a resolution at the earth's surface of 25 feet or better with a location accuracy objective of plus or minus one mile; maximum possible ground coverage; and recovery of latent image film "by means of ballistic re-entry and land or sea recovery." After identifying the primary subcontractors and items that the government would furnish, the statement turned to the question of managing the organizational amalgam. Overall technical direction, it advised the firm, "is the joint responsibility of several agencies of the Government. In the interest of effective management, however, such direction will be provided primarily by and through the Air Force Ballistic Missile Division acting as the agent for all interested components."[99]

At the end of April 1958, CORONA participants thought that they had embarked on a short-term, high-risk strategic reconnaissance venture that would augment the U-2 as an overhead technical collection system until WS-117L satellites became operational in 1961. That CORONA would succeed beyond anyone's expectations, that it would eclipse the WS-117L program entirely, that it would continue in operation over 12 years and set the pattern for American reconnaissance satellite projects to follow, and that managing it would prompt creation of a National Reconnaissance Office, they could not know and would not have believed. On the recommendation of science advisors, on the approval of the president, on the word of businessmen and government officials pledged in the clasping of hands, and on a broadly drawn two-page Statement of Work, Project CORONA was underway. The space-based "intelligence revolution" had begun.

DWAYNE A. DAY

5 A STRATEGY FOR RECONNAISSANCE

Dwight D. Eisenhower and Freedom of Space

Winston Churchill once characterized the Soviet Union as a "riddle wrapped in a mystery inside an enigma." Other scholars have used the analogy of Russian *matrioshka* dolls—open one and you will find another, and another inside that, and so on. The meaning is the same: it was difficult to know what was happening within the Soviet Union because so much was hidden behind several layers of secrecy. Therefore, the United States after World War II developed a vast intelligence bureaucracy, and the Soviet Union was its primary target.

The same analogy of secrecy could also be used regarding early American space policy under President Dwight D. Eisenhower. Rather than providing only an uncoordinated, retroactive response to Soviet events, as was commonly believed, early American space policy was highly focused, with one goal in mind: the eventual deployment of intelligence collection satellites.

SPACE AND NATIONAL SECURITY POLICY

Presidential policy is not created all at once, but rather evolves over time. As such, it is never possible to state with authority that the president's apparent goals during the implementation of a program or policy are the same ones he had at its initiation. Nevertheless, it is reasonable to conclude that Dwight D. Eisenhower, from the first time he gave serious consideration to the concept until the day he left office in January 1961, viewed satellite reconnaissance as a precious commodity that he had to protect with "bodyguards" of cover stories, half-truths, misdirections, and diversions. Much of the American civilian space program, for instance, appears to have been a visible, public means of diverting attention from the security-related programs Eisenhower valued. Even other

President Dwight D. Eisenhower addressing the American public on July 25, 1955. Eisenhower was determined to ensure that the United States gathered the intelligence it needed on the Soviet threat, but that it did so in a non-threatening manner.

military programs themselves shielded his top priority: the reconnaissance satellite program.

The recent declassification of a large number of National Security Council documents from the mid-1950s clearly shows that a crucial starting point of the American civilian space program—the U.S. scientific satellite for the International Geophysical Year (IGY)—was in fact initiated because Eisenhower and his top advisors wanted to establish a legal precedent for flying reconnaissance satellites over the Soviet Union. Eisenhower approved the U.S. scientific satellite program in 1955 as a means of serving as a "stalking horse" for future intelligence satellite programs.[1] Furthermore, the Department of Defense was not the only agency interested in this strategy. The CIA was as well, and actually contributed several million dollars in funding to the "civilian" U.S. scientific satellite program.

THE KILLIAN REPORT

In September 1954, the Science Advisory Committee of the Office of Defense Mobilization, under orders from President Eisenhower, began a study of the problem of surprise attack.[2] Headed by James Killian, then president of the Massachusetts Institute of Technology, the group was known as the

Technological Capabilities Panel (TCP). Its report, "Meeting the Threat of Surprise Attack," was issued on February 14, 1955. Often referred to as the "Killian Report," it greatly impressed Eisenhower.

During the course of its deliberations, the study's intelligence panel, headed by Polaroid's Edwin "Din" Land, became aware of two advanced proposals for intelligence collection. One was a nuclear-powered reconnaissance satellite using a television camera. This was outlined in a report by the Air Force's think-tank, the RAND Corporation, titled "Project Feed Back." The other idea was a U.S. Air Force development program for a high-altitude reconnaissance aircraft. While investigating the aircraft program, the panel became aware of a proposal by the Lockheed Skunk Works for its own high-flying strategic reconnaissance aircraft, known as the CL-282. Land brought Lockheed's proposal to President Eisenhower's attention. Unlike the Air Force program to develop a reconnaissance aircraft, the CL-282 would be configured to carry out strategic reconnaissance prior to hostilities—pre-D-day reconnaissance. The Strategic Air Command had previously rejected this mission, but those on the TCP thought it vital.

Eisenhower approved the CL-282 project and placed it under the charge of Richard Bissell, a newcomer to the CIA. The plane, eventually known as the U-2, was never mentioned in the Killian Report itself, but was described in a highly classified annex. It was labeled for the "Eyes Only" of President Eisenhower, and he probably destroyed it along with another classified annex on the submarine-launched ballistic missile program.[3]

Eisenhower assigned the U-2 mission to the CIA for three reasons. First, he thought it would be less provocative if a civilian pilot, rather than a military one, flew the aircraft into foreign territory. Second, he wanted the reconnaissance photographs to be evaluated at the national leadership level, as opposed to being evaluated within the military services (which he felt had an incentive to interpret intelligence in their own favor). Finally, he did not want to antagonize the Soviets

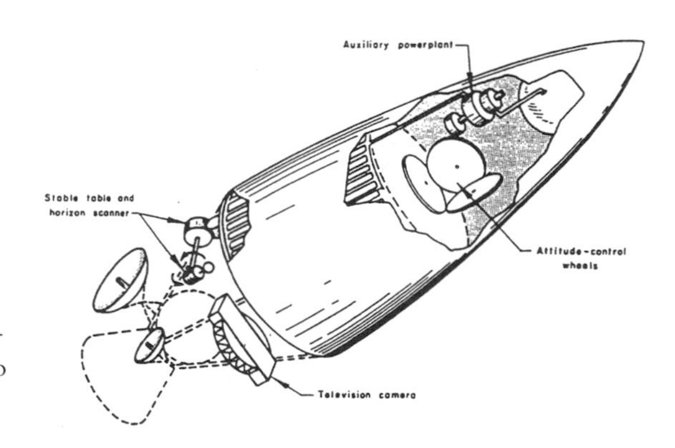

A drawing of an atomic-powered reconnaissance satellite proposed by the Project Feed Back team. This proposal prompted the Technological Capabilities Panel to recommend that the United States use a scientific satellite to establish the right to "freedom of space."

by pursuing a provocative program in the open. He felt that military pursuit of the program would only exacerbate tensions between the superpowers.[4]

Those involved in the TCP report knew that overflight of another nation's territory by such an aircraft was a clear violation of international law. It could also be viewed as a hostile act. In reality, American aircraft flying missions on the periphery of the Soviet Union were being fired upon on a regular basis and even shot down. The consequences of violating national airspace were clearly a major concern for those planning aircraft reconnaissance missions. But a satellite, flying much higher, would not necessarily violate international law since no clear definition existed of where "airspace" ended and "space" began. Realizing this, Land and the others on the panel decided to attempt to influence strongly the *establishment* of international law.

Although the majority of the contents of the intelligence section of the TCP report remains classified, crucial portions are now known because they were included in other documents that have been declassified. For example, a cover letter accompanying the report upon its delivery to the Policy Planning Staff at the Department of State declares that one of the intelligence panel's conclusions was the need for the "re-examination of the principles of freedom of space, particularly in connection with the possibility of launching an artificial satellite into an orbit about the earth, in anticipation of use of larger satellites for intelligence purposes."[5] Another document, dated only two weeks later, called for the Department of State to discuss the following recommendation:

> Freedom of Space. The present possibility of launching a small artificial satellite into an orbit about the earth presents an early opportunity to establish a precedent for distinguishing between "national air" and "international space," a distinction which could be to our advantage at some future date when we might employ larger satellites for intelligence purposes.[6]

Land, Killian, and others considered a reconnaissance satellite to be technologically unrealistic in the near future. They believed that the CL-282 and an Air Force reconnaissance balloon program known as GENETRIX were more realistic near-term possibilities. But they also felt that the United States should begin establishing a precedent to enable future satellite reconnaissance missions. The TCP advocated such a strategy.

THE SCIENTIFIC SATELLITE PROGRAM

In parallel to the deliberations of the TCP, serious proposals for an initial U.S. scientific satellite emerged. Wernher von Braun and his colleagues at the Army

Ballistic Missile Agency in Huntsville, Alabama, teamed up with the Office of Naval Research to propose a satellite called Orbiter. Later in the year, the American Rocket Society prepared a detailed survey of possible scientific and other uses of a satellite and proposed to the National Academy of Sciences' U.S. National Committee for the IGY that such a satellite be developed for use during the IGY.

The Feed Back and the Killian reports were both highly secret, although the Orbiter proposal was not. The CL-282 proposal, in particular, was known to only a handful of people. One person who knew of all three projects, as well as the TCP report, was Assistant Secretary of Defense for Research and Development Donald Quarles, no stranger to proposals for earth-orbiting satellites. In 1952, President Truman had asked Aristid Grosse, a physicist at Temple University, to prepare a report on the "Present Status of the Satellite Problem." The report was not completed before Truman left office and instead was presented to Quarles, who was in charge of virtually all defense-related research projects in the early years of the Eisenhower presidency.

On March 14, 1955 (the same day the TCP report was completed), the U.S. National Committee on the International Geophysical Year of the National Academy of Sciences (NAS) presented a recommendation to Alan Waterman, the director of the National Science Foundation (NSF). The committee recommended that a scientific satellite be developed as part of the IGY.[7] Quarles, who was aware of the TCP report's findings, asked Waterman to formally suggest this idea to the National Security Council. Four days later Waterman sent a letter to Deputy Undersecretary of State Robert Murphy, proposing that the United States conduct just such a scientific mission.[8] Those at the National Academy would not have been privy to information about Feed Back, CL-282, or the TCP report, so they were unlikely to have recognized the strategic objectives behind the satellite proposal.

Four days later Murphy met with Waterman, Detlev Bronk, president of the NAS, and Lloyd Berkner, one of the academy's influential members, to discuss the issue. In a letter one month later, Murphy stated that such a proposal would "as a matter of fact, undoubtedly add to the scientific prestige of the United States, and it would have a considerable propaganda value in the cold war."[9] Having gained the support of the Department of State, Waterman then discussed the issue once again with Quarles, who suggested that he consult the Director of Central Intelligence, Allen Dulles, on how best to proceed. Waterman did so and gained Dulles's support for the program. He also spoke with Percival Brundage, the director of the Bureau of the Budget, to gain his cooperation when needed. Thus the scientific satellite proposal now had the support of the departments of State and Defense and the CIA, as well the Bureau of the

Budget. Waterman also agreed to formally propose the full program to an executive session of the National Science Board on May 20.[10]

NSC 5520

The speed at which these events took place is startling. Only nine weeks after the TCP report recommended the approval of a scientific satellite program, on May 20, 1955, the National Security Council approved a top-level policy document known as NSC 5520, "U.S. Scientific Satellite Program." The document stated that the United States should develop a small scientific satellite weighing 5–10 pounds. It stated:

> The report of the Technological Capabilities Panel of the President's Science Advisory Committee recommended *that intelligence applications warrant* an immediate program leading to a very small satellite in orbit around the earth, and that re-examination should be made of the principles or practices of international law with regard to "Freedom of Space" from the standpoint of recent advances in weapon technology.[11]

The document continued:

> From a military standpoint, the Joint Chiefs of Staff have stated their belief that intelligence applications strongly warrant the construction of a large surveillance satellite. While a small scientific satellite cannot carry surveillance equipment and therefore will have no direct intelligence potential, it does represent a technological step toward the achievement of the large surveillance satellite, and will be helpful to this end so long as the small scientific satellite program does not impede development of the large surveillance satellite.[12]

NSC 5520 also stated:

> Furthermore, a small scientific satellite will provide a test of the principle of "Freedom of Space." The implications of this principle are being studied within the Executive Branch. However, preliminary studies indicate that there is no obstacle under international law to the launching of such a satellite.
>
> It should be emphasized that a satellite would constitute no active military offensive threat to any country over which it might pass. Although a large satellite might conceivably serve to launch a guided missile at a ground target, it will always be a poor choice for the purpose. A bomb could not be dropped from a satellite on a target below, because anything dropped from a satellite would simply continue alongside in the orbit.

Although the document correctly noted the limited utility of satellites as active military offensive threats, this was not the purpose of the surveillance satellite program. In fact, deploying weapons systems in orbit has never been a major aspect of American military space policy.

A year later, another White House document stated that the scientific satellite program would have the target date of 1958 "with the understanding that the program developed thereunder will not be allowed to interfere with the ICBM and IRBM programs but will be given sufficient priority by the Department of Defense in relation to other weapons systems to achieve the objectives of NSC 5520."[13] In other words, the scientific satellite would not be allowed to interfere with ballistic missile programs like Atlas and Jupiter, but it also would not be allowed to languish without attention while the Department of Defense focused on other programs such as military satellites. Such inattention would have effectively undercut the entire purpose of NSC 5520, to lay the foundation for reconnaissance satellites. At the same time, the launching of such a satellite was *not* to interfere with the development of those satellites. In reality, this latter point was moot, for the reconnaissance satellite program was severely underfunded at the time.

NSC 5520 directed that the scientific satellite program be associated with the International Geophysical Year. "Freedom of space" could be challenged by the Soviets more easily if the United States simply launched a satellite at any time. The IGY provided an international context for conducting the program—a legitimate cover story even better than the civilian auspices of the existing program. Even if the Soviets had not announced their intention to build a satellite for the IGY, the United States would be able to use the IGY to justify its program if the Soviets complained. Quarles himself specifically noted this only a week after the approval of NSC 5520.[14]

WAS A POLICY EVEN NEEDED?

The TCP report, perhaps one of the most influential documents of the early Cold War, served as the starting point for a number of major American defense programs in the 1950s. It recommended the development of not only what became the U-2, the scientific satellite program, and reconnaissance satellites, but also what became the Polaris submarine-launched ballistic missile, the SOSUS underwater sonar array, and the extension of the Distant Early Warning (DEW) Line.

Implementing this extensive list of recommendations required a concerted effort by various agencies and military services. It required an extensive oversight effort as well, a task that fell to the Operations Coordinating Board of the National Security Council, which produced a progress report in June 1955 on the status of the TCP recommendations. This progress report, known as NSC

5522, also included various government departments' views on the panel's recommendations.

In response to the TCP recommendation concerning freedom of space, NSC 5522 stated:

> State, Treasury, Defense and Justice Comment: Any unilateral statement by the U.S. concerning the freedom of outer space is unnecessary. It is clear that the jurisdiction of a state over the air space above its territory is limited and that the operation of an artificial satellite in outer space would not be in violation of international law. State and Justice point out that by the convention on international civil aviation of 1944 (to which the U.S. is a party, but the U.S.S.R. is not) and by customary law every State has exclusive sovereignty "over the air space above its territory." However, air space ends with the atmosphere. There has been no recognition that sovereignty extends into airless space beyond the atmosphere.[15]

This statement may have appeared self-evident to those involved, but in effect it begged the question: if space was not sovereign territory, how was a nation to determine where the atmosphere ended and space began? The edge of the atmosphere was not a clearly defined boundary and, indeed, could not be defined until satellites began orbiting the earth to measure the extent of the atmosphere. Still, it made sense for the departments to advise that the United States not issue a unilateral statement concerning freedom of space, for such a statement would only draw attention to the subject at a time when the whole point of the scientific satellite program was to establish freedom of space in as innocuous a manner as possible.

The CIA also weighed in on the matter. Its comments are very illuminating and highly prescient, especially in light of the world reaction to Sputnik two years later:

> The psychological warfare value of launching the first earth satellite makes its prompt development of great interest to the intelligence community and may make it a crucial event in sustaining the international prestige of the United States.
>
> There is an increasing amount of evidence that the Soviet Union is placing more and more emphasis on the successful launching of the satellite. Press and radio statements since September 1954 have indicated a growing scientific effort directed toward the successful launching of the first satellite. Evidently the Soviet Union has concluded that their satellite program can contribute enough prestige of cold war value or knowledge of military value to justify the diversion of the necessary skills, scarce material and labor from immediate military production. If the Soviet effort should prove successful before a similar United States effort, there is no doubt but that their propaganda would capitalize on the theme of the scientific and industrial superiority of the communist system.

The successful launching of the first satellite will undoubtedly be an event comparable to the first successful release of nuclear energy in the world's scientific community, and will undoubtedly receive comparable publicity throughout the world. Public opinion in both neutral and allied states will be centered on the satellite's development. For centuries scientists and laymen have dreamed of exploring outer space. The first successful penetration of space will probably be the small satellite vehicle recommended by the Technological Capabilities Panel. The nation that first accomplishes this feat will gain incalculable prestige and recognition throughout the world.

The United States' reputation as the scientific and industrial leader of the world has been of immeasurable value in competing against Soviet aims in both neutral and allied states. Since the war the reputation of the United States' scientific community has been sharply challenged by Soviet progress and claims. There is little doubt but that the Soviet Union would like to surpass our scientific and industrial reputation in order to further her influence over neutralist states and to shake the confidence of states allied with the United States. If the Soviet Union's scientists, technicians and industrialists were apparently to surpass the United States and first explore outer space, her propaganda machine would have sensational and convincing evidence of Soviet superiority.

If the United States successfully launches the first satellite, it is most important that this be done with unquestionable peaceful intent. The Soviet Union will undoubtedly attempt to attach hostile motivation to this development in order to cover her own inability to win this race. To maximize the effectiveness of Soviet accusations, the satellite should be launched in an atmosphere of international good will and common scientific interest. For this reason, the CIA strongly concurs in the Department of Defense's suggestion that a civilian agency such as the U.S. National Committee of the IGY supervise its development and that an effort be made to release some of the knowledge to the international scientific community.

The small scientific vehicle is also a necessary step in the development of a larger satellite that could possibly provide early warning information through continuous electronic and photographic surveillance of the USSR. A future satellite that could directly collect intelligence data would be of great interest to the intelligence community.

The Department of Defense has consulted with the [Central Intelligence] Agency, and we are aware of their recommendations, which have our full concurrence and strong support.[16]

LEGAL ISSUES

If those involved with the legal issues needed their thinking focused about the subject of international airspace, they could ask for no better lesson than that provided by the GENETRIX reconnaissance balloon program. Beginning in January 1956 hundreds of balloons were launched to drift over the Soviet Union, their cameras photographing the countryside. The majority were never seen again and American officials knew that they had come down (or been shot down) inside the Soviet Union. On February 7, anticipating the Soviet response, Eisenhower suggested to Secretary of State John Foster Dulles that the operation be suspended and "we should handle it so it would not look as though we had been caught with jam on our fingers." On February 9 the Soviets held a press conference outside Spridonovka Palace. About fifty balloons and instrument containers were placed on display. The balloons, the Soviets said, were part of an espionage project and had been a clear violation of their airspace. This was a major embarrassment for the United States.

Meanwhile, legal issues concerning exactly what "freedom of space" meant occupied U.S. government lawyers. The State Department's Policy Planning Staff was assigned the task of reporting on-going progress on achieving several of the TCP report's recommendations. On October 2, 1956, the Policy Planning Staff reported some preliminary thinking on the issue of freedom of space:

So far as law is concerned, space beyond the earth is an uncharted region concerning which no firm rules have been established. The law on the subject will necessarily differ with the passage of time and with practical efforts at space navigation. Various theo-

A downed GENETRIX reconnaissance balloon system recovered in China. Hundreds of GENETRIX balloons and their camera payloads came down in the Soviet Union and China in 1956 and embarrassed the United States. President Eisenhower finally forbade their future use. (Photo courtesy of Dino Brugioni)

ries have been advanced concerning the upper limits of a state's jurisdiction, but no firm conclusions are now possible.

A few tentative observations may be made: (1) A state could scarcely claim territorial sovereignty at altitudes where orbital velocity of an object is practicable (perhaps in the neighborhood of 200 miles); (2) a state would, however, be on strong ground in claiming territorial sovereignty up through the "air space" (perhaps ultimately to be fixed somewhere in the neighborhood of 40 miles); (3) regions of space which are eventually established to be free for navigation without regard to territorial jurisdiction will be open not only to one country or a few, but to all; (4) if, contrary to planning and expectation, a satellite launched from the earth should not be consumed upon reentering the atmosphere, and should fall to the earth and do damage, the question of liability on the part of the launching authority would arise.[17]

SELECTING A SATELLITE PLAN

After the scientific satellite concept was approved by NSC 5520, it had to be turned into programmatic reality. Not surprisingly, this task was given to Donald Quarles, who was essentially the adminstration's point man on space issues. On June 8, 1955, Secretary of Defense Charles Wilson delegated to Quarles the responsibility for implementing the plan, which had essentially been Quarles's idea all along. Quarles then created an Advisory Group on Special Capabilities, chaired by Homer J. Stewart, to select a scientific satellite and a method for launching it. The advisory group held hearings, listened to presentations by the Army, Navy, and Air Force, and ultimately selected the Navy's proposal, Project Vanguard. The Air Force, already involved in the development of an intelligence satellite, was not seriously interested in the scientific satellite program.[18]

Many of those involved, particularly Wernher von Braun and his group of engineers working under Army auspices (for the Army Ballistic Missile Agency, or ABMA), were surprised that the Army team had lost out, since they had begun developing their proposal a year before. Years later, when explaining the decision, members of the advisory group gave a number of reasons, including Navy developments in satellite instrumentation, the supposed lower costs and growth potential of the Vanguard, and the desire to avoid using a ballistic missile for the program.[19] The last point was the most relevant one: NSC 5520 was quite explicit in stating that the satellite program should not interfere at all with ICBM and IRBM programs. There was no way that the Army proposal could be selected without interfering with Army development of the Redstone booster as the basis of a ballistic missile to be called Jupiter—at least without significant additional cost.

THE DECISION IS REEVALUATED

The initial decision in favor of the Navy program did not settle the issue. Von Braun's group justifiably felt that they were more qualified to complete the mission than the unproven Vanguard team. Throughout 1955 and 1956 the Army team proceeded with development of a Jupiter/Redstone rocket designed to conduct re-entry tests of ballistic nose cones. It also continued to lobby behind the scenes to obtain permission to proceed with its own satellite launching program.

Von Braun's group's complaints about not being selected for the program eventually reached Washington. On April 23, 1956, the Army informed the Office of the Secretary of Defense that a Jupiter could orbit a satellite in January 1957. On May 16, 1956, the Secretary of Defense's Special Assistant for Guided Missiles rejected any use of the Jupiter as an alternative to Vanguard. Apparently not convinced that the Army had gotten the message, six days later the Office of the Assistant Secretary of Defense informed the ABMA that no plans or preparations should be initiated for using the Redstone or Jupiter as launch vehicles.

But in government there is never such a thing as a final answer. So ABMA took its appeal to the National Security Council. It complained to Deputy Secretary of State Herbert Hoover Jr., who was chairman of the Operations Coordinating Board, assigned to monitor NSC directives. Colonel Andrew Goodpaster, Eisenhower's staff secretary, reported the group's appeal:

> On May 28th Secretary [Deputy Secretary of State] Hoover called me over to mention a report he had received from a former associate in the engineering and development field regarding the earth satellite project. The best estimate is that the present project would not be ready until the end of '57 at the earliest, and probably well into '58. Redstone had a project well advanced when the new one was set up. At minimal expense ($2–$5 million) they could have a satellite ready for firing by the end of 1956 or January 1957. The Redstone project is one essentially of German scientists and it is American envy of them that has led to a duplicative project.
>
> I spoke to the President about this to see what would be the best way to act on the matter. He asked me to talk to Secretary Wilson. In the latter's absence, I talked to Secretary [Deputy Secretary of Defense Reuben B.] Robertson today and he said he would go into the matter fully and carefully to try to ascertain the facts. In order to establish the substance of this report, I told him it came through Mr. Hoover (Mr. Hoover had said I might do so if I felt it necessary).[20]

Von Braun's group, not privy to the strategic considerations that had led to its losing the competition, blamed the decision on anti-German prejudice. This incident started a new evaluation of the satellite selection process, which ultimately rejected the Huntsville group once again. On June 22, 1956, Homer J.

Stewart reported the results of two meetings held by the group in late April to Charles C. Furnas, Assistant Secretary of Defense for Research and Development (by this time Quarles had left that position to become Secretary of the Air Force). Stewart reported that although Project Vanguard was suffering some minor setbacks and was short of highly capable people, in general the project was on a satisfactory schedule and "one or more scientific satellites can be successfully placed in orbit during the IGY."[21] Stewart also stated:

> Redstone re-entry vehicle No. 29, now scheduled for firing in January 1957, apparently will be technically capable of placing a 17 pound payload consisting principally of radio beacons and doppler-equipment in a 200-mile orbit, even with the degradation in performance below the present design figures which might reasonably be expected, but without any appreciable further margin. This capability will depend upon successful accomplishment of several developments, such as the use of a new fuel in the Redstone booster, and the spinning cluster of fifteen solid propellant motors. The probability of success of this single flight cannot be reliably predicted now, but it would doubtless be less than 50 per cent.

Stewart explained why the Army proposal should be rejected once again:

> In any case, such a single flight would not fulfill the Nation's commitment for the International Geophysical Year because it would have to be made before the beginning of that period. Adequate tracking and observation equipment for the scientific utilization of results would not be available at this time. Moreover, any announcement of such a flight (or worse, any leakage of information if no prior announcement were made) would seriously compromise the strong moral position internationally which the United States presently holds in the IGY due to its past frank and open acts and announcements as respects VANGUARD.

Stewart mentioned that a Redstone (in actuality a Jupiter) could be used as a backup later in 1957 if Vanguard fell behind schedule and that No. 29 was the only vehicle that could be used without interfering with the Redstone program. He concluded, "At the present time, therefore, the Group does not recommend activating a satellite program based on the Redstone missile, but will reconsider this question and the possibilities of the ICBM program at its subsequent meetings when the critical items of the VANGUARD program are further advanced."[22]

On July 5, 1956, E. V. Murphree, DoD's Special Assistant for Guided Missiles, wrote to Reuben B. Robertson Jr., the Deputy Secretary of Defense. Murphree stated that he had looked into the possibility of using a Jupiter reentry test vehicle for launching a satellite into orbit. He stated that the January 1957 test could be adapted to this purpose with little effort and no impact on the program and said that an attempt could be made in September 1956, although this *would* affect the Jupiter program.

Murphree further noted that proposals for using the Jupiter were not new and that the original Redstone satellite and reentry test vehicle proposals resulted from a common study (made by Wernher von Braun) which argued that the same vehicle could be used for both. Murphree also stated that the first two tests of the Jupiter were essentially propulsion system tests and could accomplish much of their goals for that program even if used for satellite launch attempts. Murphree continued, "There is, however, room for serious doubt that two isolated flight attempts would result in achieving a successful satellite, and the dates of such flights would be *prior* to the Geophysical Year for which a satellite capability is specifically required, and prior to the time when tracking instrumentation will be available."[23]

Murphree then stated that these facts had been taken into consideration at the time that the Office of the Secretary of Defense reviewed the satellite program and decided to assign the mission to the Navy group. He said, "That decision was based largely on a conviction that the VANGUARD proposal offered the greater promise for success. The history of increasing demands for funds for this program confirms the conviction that this is not a simple matter. I know of no new evidence available to warrant a change in that decision at this time."

The Vanguard satellite was small, scientific, and, most important, civilian—all necessary for peacefully establishing the right of "freedom of space." Here a Vanguard satellite sits atop its launch vehicle.

The rest of Murphree's memorandum is extremely interesting and is reprinted below in full:

While it is true that the VANGUARD group does not expect to make its first satellite attempt before August 1957, whereas a satellite attempt could be made by the Army Ballistic Missile Agency as early as January 1957, little would be gained by making such an early satellite attempt as an isolated action with no follow-up program. In the case of VANGUARD, the first flight will be followed up by five additional satellite attempts in the ensuing year. It would be impossible for the ABMA group to make any satellite attempt that has a reasonable chance of success without diversion of the efforts of their top-flight scientific personnel from the main course of the JUPITER program, and to some extent, diversion of missiles from the early phase of the re-entry test program. There would also be a problem of additional funding not now provided.

For these reasons, I believe that to attempt a satellite flight with the JUPITER re-entry test vehicle without a preliminary program assuring a very strong probability of its success would most surely flirt with failure. Such probability could only be achieved through the application of a considerable scientific effort at ABMA. The obvious interference with the progress of the JUPITER program would certainly present a strong argument against such diversion of scientific effort.

On discussing the possible use of the JUPITER re-entry test vehicle to launch a satellite with Dr. Furnas, he pointed out certain objections to such a procedure. He felt there would be a serious morale effect on the VANGUARD group to whom the satellite test has been assigned. Dr. Furnas also pointed out that a satellite effort using the JUPITER re-entry test vehicle may have the effect of disrupting our relations with the non-military scientific community and international elements of the IGY group.

I don't know if I have a clear picture of the reasons for your interest in the possibility of using the JUPITER re-entry test vehicle for launching the satellite. I think it may be helpful if Dr. Furnas and I discuss this matter with you, and I'm trying to arrange for a date to do this on Monday.[24]

Robertson's special assistant, Charles G. Ellington, forwarded the memos to White House Staff Secretary Andrew Goodpaster. Goodpaster wrote on the bottom of Ellington's cover memo, "Secy. Robertson feels no change should be made—per Mr. Ellington. Reported to President."[25]

HOW TO "BEAT" THE SOVIETS

As one would expect, scientists involved in the satellite program stressed that a scientifically useful program would enhance U.S. prestige. National Science Foundation Director Alan Waterman specifically noted that the schedule was less important than the prestige to be gained from a program that produced a major scientific breakthrough. In other words, being first but not scientifically significant was not as important for national prestige as accomplishing something scientifically noteworthy.[26] One could win the physical race, but lose the scientific competition.

Rather surprisingly, this was also the conclusion of the National Security Council's Planning Board. NSC 5520 had mentioned the possibility of the Soviets developing a satellite, but took no position on whether the United States should attempt to launch before the Soviets. The Defense Department's response in NSC 5522 stated with more urgency that the United States should be first, but must do so carefully. The NSC Planning Board, in November 1956, made this remarkable statement:

> The USSR can be expected to attempt to launch its satellite before ours and to attempt to surpass our effort in every way. It is vitally important in terms of the stated prestige and psychological purposes that the United States make every effort to (1) make possible a successful launching as soon as practicable and (2) put on as effective an IGY scientific program as possible. The prestige and psychological set-backs inherent in a possible earlier success and larger satellite by the USSR could at least be partially be offset by a more effective and complete scientific program by the United States. Even if the United States achieved the first successful launching and orbit, but the USSR put on a stronger scientific program, the United States could lose its initial advantage.[27]

Thus, one year before Sputnik, the NSC's official position was that the United States could lose prestige *even* if it launched a satellite first but the Soviets developed a better space science program. Therefore, science was given higher priority than schedule.

The issue of the propaganda value of launching a satellite had been mentioned in numerous documents, including NSC 5520 and NSC 5522.[28] It was even the focus of a study at the RAND Corporation. Clearly, many top U.S. policymakers felt that it was a major issue. Eisenhower always dismissed these concerns. He did not believe that it would be that important an issue.[29] From a domestic political standpoint, he proved to be dreadfully wrong, but not necessarily from the standpoint of international law.

SPUTNIK CRISIS?

By the fall of 1957, the Vanguard satellite program was proceeding roughly on schedule, with first launch anticipated for late 1957 or early 1958, during the IGY. No one expected the first launch to be successful. The satellite reconnaissance program, however, was underfunded, and not making significant progress. Land and Killian took renewed interest in it by the fall of 1957 and the Science Advisory Committee even sponsored a special briefing for the White House on the subject on September 20.[30] The strategy was proceeding, but by October it was overtaken by events.

On October 4, 1957, the Soviet Union launched a 184-pound metallic ball called Sputnik. It was instantly major news around the world. On October 7, acting on the publicity generated in the wake of Sputnik, President Eisenhower asked Quarles, by then Deputy Secretary of Defense, to explain why the United States was in the position it was. Goodpaster related Quarles's explanation in the minutes of the meeting:

> The Science Advisory Committee had felt, however, that it was better to have the Earth satellite proceed separately from military development. One reason was to stress the peaceful character of the effort, and a second was to avoid the inclusion of material, to which foreign scientists might be given access, which is used in our own military rockets. . . . [Quarles] went on to add that the Russians have in fact done us a good turn, unintentionally, in establishing the concept of freedom of international space—this seems to be generally accepted as orbital space, in which the missile is making an inoffensive passage.[31]

Two days later, Eisenhower mentioned this issue again. "When military people begin to talk about this matter, and to assert that other missiles could have been used to launch a satellite sooner, they tend to make the matter look like a 'race,' which is exactly the wrong impression."[32] Eisenhower's fiscal conservatism was yet another important part of this deliberate strategy—he did not want to spend a lot of money in an arms race or a space race with the Soviets. But equally important was his desire to not unnecessarily inflame the Cold War. Interservice rivalry was not simply expensive; it also exacerbated tensions with the Soviet Union.

THE CIA AND THE SCIENTIFIC SATELLITE

When Alan Waterman met with Allen Dulles in early May 1955 to discuss NSF sponsorship of a scientific satellite, he also met with Richard Bissell, whom he described as "the one in Central Intelligence who is following this closely."[33] Bissell had good reasons to follow it closely, since he was then in the middle of managing the newly created U-2 reconnaissance aircraft program. Use of the U-2 would violate international law.

As the scientific satellite program continued and ran into significant cost overruns, surprisingly the CIA provided money for it to continue. In April 1957, the director of the Bureau of the Budget, Percival Brundage, sent a lengthy memo to Eisenhower on Vanguard cost overruns. Vanguard was initially supposed to cost $15–20 million. By spring 1957, it was projected to cost 10 times that amount. Brundage recounted the funding difficulties of the program and stated, "Apparently, both the Department of Defense and the National Science Foundation are very reluctant to continue to finance this project to completion. But each is quite prepared to have the other do so." Brundage also noted that the National Science Foundation had contributed an extra $5.8 million in funds to the Department of Defense to fund the program and that the CIA had contributed $2.5 million of its own money as well.[34]

Why was the CIA, which had no official stake in the scientific satellite program and no space programs underway or even under study, willing to spend its own money on a civilian scientific space program? The answer is unknown. But it was likely due to Bissell's intervention, since he was the person delegated to follow the program and he was also in charge of the U-2 reconnaissance aircraft.[35] By April 1957, when the CIA provided the money to the scientific satellite program, the U-2 had already made nearly a dozen flights over the Soviet Union, each protested vigorously but quietly by the Soviets. Although the CIA did not have a reconnaissance satellite program at this time, it would be given one by Eisenhower less than a year later, a program code-named CORONA. Bissell was placed in charge of that endeavor as well. Bissell was in control of a substantial amount of funding for covert operations at CIA and it is likely that the money to support the scientific satellite program came from these funds or from DCI Allen Dulles's substantial discretionary account.

Brundage concluded his memo to Eisenhower by noting that the Air Force had already started its own, much larger reconnaissance satellite. "Therefore, whether or not the International Geophysical Year satellite project is completed, research in this area will not be dropped." But this missed the point, since the scientific satellite program had less to do with research than with establishing

Richard Bissell, as CIA manager of the U-2 program, provided $2.5 million of CIA funding for the U.S. scientific satellite program. He wanted to ensure that the legal questions that surrounded the U-2 aircraft's overflight of foreign territory did not apply to reconnaissance satellites. (Photo courtesy of the CIA)

legal precedent. Brundage may have been yet another top government official unaware of this secret strategy.

GOING UNDER COVER

The Soviets' launch of Sputnik II on November 5 had an even more profound public relations impact than its predecessor. Sputnik II was not only far larger than Sputnik I, but it also carried a dog, demonstrating a sophistication that belied early administration attempts to downplay the Soviet achievement. The size of the payload was clearly sufficient for the carrying of an atomic bomb, which heightened Americans' fear that the Soviets could now effectively attack the United States. A little over a month later, the U.S. attempt to launch the Vanguard satellite ended in embarrassing failure. Clearly, in the realm of space exploration, the Soviets had taken a substantial lead. From an international legal standpoint, this should have rendered the subject of freedom of space moot, but it did not.

In November, newly appointed Secretary of Defense Neil McElroy proposed centralizing control of the various American space projects then underway, such as Vanguard and WS-117L, along with advanced ballistic missile development. They would be put in a Defense Special Projects Agency (DSPA), which would be responsible for whatever projects the secretary would assign to it. The idea for the DSPA apparently came from the Science Advisory Council in mid-October, just days after both Sputnik's launch and McElroy's nomination.[36] Eisenhower himself expressed the opinion that a fourth service should be established to handle the "missiles activity."[37] McElroy said that he was weighing the idea of a "Manhattan Project" for antiballistic missiles. The president thought that a separate organization might be a good idea for this problem as well.[38] In testimony before Congress, Quarles, who might easily have been regarded as an Air Force partisan, stated that long-range, surface-to-surface missiles had been assigned to the Air Force because it possessed the targeting and reconnaissance capabilities to use them, not because they accomplished a unique Air Force mission.[39] Space could conceivably be treated in the same way.

Although the plan for incorporating ballistic missile development in it was eliminated, the idea of a new space agency proceeded. A Special Projects Agency would act as a central authority for all U.S. space programs and would essentially contract out missions to the separate services, civilian government agencies, and even universities and private industry. "Above the level of the three military services, having its own budget, it would be able to concentrate on the new and the unknown without involvement in immediate requirements and inter service rivalries." McElroy stated in front of Congress that "the vast

A mockup of the Sputnik II payload complete with Soviet dog. When the Soviet Union launched Sputniks I and II, they raised fears of surprise attack among many in the West. But they also unwittingly played into the Eisenhower administration's hands and established the concept of "freedom of space" for future reconnaissance satellites. (Photo courtesy of the National Reconnaissance Office)

weapons systems of the future in our judgment need to be the responsibility of a separate part of the Defense Department."[40] This proposal was placed in a DoD reorganization bill. At this point, it was still assumed that the entire American space program would remain under military control, although at the level of the Secretary of Defense in an office specially created to manage it.

Discussion of the Defense Special Projects Agency continued within the administration. Its name was changed to the Advanced Research Projects Agency (ARPA) and Eisenhower sent a message to Congress on January 7 requesting supplemental appropriations for the agency.[41]

On February 7, 1958, James Killian and Din Land, who was also a member of the President's Board of Consultants on Foreign Intelligence Activities, met with Eisenhower and his staff secretary, General Andrew Goodpaster. They briefed the president on the potential of both a recoverable space capsule and a supersonic reconnaissance aircraft program, suggesting that in order to speed up the development of a reconnaissance satellite, the United States should pursue the recoverable capsule idea as an "interim" solution. Eisenhower accepted this recommendation and the satellite program was soon named CORONA.

An equally important result of this first meeting was the decision to finalize Secretary of Defense McElroy's proposal and create ARPA to house highly technical defense research programs. General Electric executive Roy Johnson was named as its director. Eisenhower decided to give ARPA control of all military space programs, including the military man-in-space program, meteorological programs, and WS-117L.

Sputnik also led to the creation of NASA, a civilian space program. Although Eisenhower initially felt that the military could handle the task of space science, he was eventually persuaded that a civilian space agency was needed, in part by Vice-President Nixon, who argued that a civilian agency was important for international prestige purposes. A military space agency could not be used as an effective means to show the flag, particularly if the president wanted to prevent escalation of the Cold War.

The creation of these new organizations, both civilian and military, proceeded apace, drawing much attention from the Congress and the public. ARPA was officially in charge of the nation's WS-117L reconnaissance satellite program, whose existence the Air Force had leaked to the press in the aftermath of Sputnik. At the same time, Eisenhower directed that the recoverable satellite program—the program that eventually became known as CORONA—be handled by a covert team involving the Air Force and the CIA. For many who had been involved in it, including those who had proposed it, the recoverable satellite program simply ceased to exist. As far as they knew, WS-117L was the only U.S. reconnaissance satellite program.

Those in charge of WS-117L eventually divided it into several programs, including the SENTRY reconnaissance program (later SAMOS), an early warning satellite known as MIDAS, and Discoverer, an engineering and development program that was to include the launching of biomedical payloads such as mice and monkeys. In reality, Discoverer was nothing more than a cover for the CORONA program, and SENTRY appears to have been continued publicly at least in part as a sleight of hand to distract attention away from Discoverer. Canceling it would have been too suspicious and would also have raised the ire of the Air Force.

FREEDOM OF SPACE MARCHES ON

The National Security Council addressed the issue of the militarization of space in August 1958. Known as NSC 5814/1, the document stated that the United States should urgently seek a political framework that would place the uses of U.S. reconnaissance satellites in a political and psychological context favorable to the U.S. intelligence effort. Responding to this, the State Department declared that one of the priorities for the United States was establishing an acceptable policy framework for the WS-117L program.[42] Even within this highly classified document, CORONA was nowhere mentioned.

Protection of WS-117L became a U.S. goal. The State Department debated satellite reconnaissance internally and eventually brought it before the United Nations. At the suggestion of the United States, the United Nations created the UN Ad Hoc Committee on the Peaceful Uses of Outer Space (COPUOS). COPUOS became the source of much heated debate over the next several years, as the Soviet Union refused to participate and instead complained about American nuclear weapons and overseas bases, stating that U.S. concessions on these issues were a prerequisite to its participation in COPUOS. Calls by COPUOS for cooperation between the superpowers on space projects were met with derision from the Soviets, who feared that such cooperative efforts would reveal the limitations of their ICBM technology. Much controversy was generated, but no actual policy evolved. While the issue was being discussed in the United Nations, the newspapers, and in classified State Department meetings, CORONA continued its development totally unknown to even those at high levels of the government who were discussing the legal protection of the more overt WS-117L program.

The advent of military reconnaissance satellites themselves created its own legal issues. In January 1959, O. G. Villard Jr. of Stanford University, a member of the National Academy of Sciences and the Space Science Board (SSB), wrote Lloyd V. Berkner, chairman of the SSB. Villard expressed concern that the U.S. military might attempt to portray its satellite launchings as scientific in nature. This deception, he suggested, could have negative effects on American space science, particularly with regard to international cooperation. Villard stated that it was in the best interests of U.S. scientists that such deception not occur.

On January 28, 1959, Berkner brought the issue to the attention of Killian, Eisenhower's science advisor. On February 13 Killian mentioned it to NASA Administrator T. Keith Glennan and Gordon Gray, Eisenhower's special assistant for national security affairs. Gray felt that the concerns raised by Villard might require further action by the National Security Council.[43] The results of their discussion of the issue are not known, but this discussion took place over a year after Eisenhower had directed that a small reconnaissance satellite program be peeled away from the bigger reconnaissance program (the one that Villard referred to in his letter) and be conducted covertly. Indeed, the first launch was scheduled to take place later in the month—under the cover of a scientific program!

The first Discoverer was launched on February 28, 1959. It was actually a test of equipment for the CORONA program. Although the satellite did not reach orbit, the Soviets still protested the flight, decrying its military nature. This prompted Richard Leghorn, the architect of Eisenhower's 1955 Open Skies proposal and the president of Itek, which was then managing the manufacture of CORONA cameras, to draft a proposal for the president titled "Political Action and Satellite Reconnaissance." In it Leghorn stated:

> The problem is not one of technology. It is not a problem of vulnerability to Soviet military measures. The problem is one of the political vulnerability of current reconnaissance satellite programs.
>
> For many years the U.S. has had overflight capabilities (aircraft and balloons) which have been substantially invulnerable to Soviet military countermeasures, but very vulnerable politically. Already the Communists (East Germany) have attacked Discoverer I as an espionage activity, and we can anticipate powerful Soviet political countermeasures to the Discoverer/SENTRY series.[44]

Leghorn continued:

> What is needed is a program to put reconnaissance satellites in the white through early and vigorous political action designed to:
>
> 1. blunt in advance Soviet political countermeasures;
> 2. gain world acceptance for the notion that the surveillance satellites are powerful servants of world peace and security, and are not illegitimate instruments of espionage;
> 3. regain the political initiative of the open skies proposal.

Leghorn was fully aware of all American reconnaissance satellite efforts and it is impossible to believe that his proposal was anything other than a continuation of his earlier thinking on overflight. But at this time the United States still had not orbited a reconnaissance satellite (and would not for over a year, suffering a string of failures with CORONA) and was not *officially* anywhere near flying a reconnaissance satellite. Taking any public position at all on the subject seemed premature at best and foolish at worst. From the White House's point of view it was best to let the political deliberations at the United Nations run their course. Leghorn's proposal went nowhere.

Gary Powers's U-2 reconnaissance aircraft was shot down on May 1, 1960, creating yet another public embarrassment for Eisenhower and revealing a carefully planned and executed strategy to gather intelligence on the Soviet Union—a strategy that also relied upon a scientific cover story to mask the true purpose of the high-flying plane. While the downing ruined the upcoming summit, the U-2 had proved an intelligence bonanza for the United States, something that Eisenhower did not wish to emphasize even in his television address to the nation following the incident.[45] Eisenhower felt that it was better to suffer the humiliation than to reveal American capabilities or gloat about American ingenuity.

Soviet Premier Khrushchev used the event to maximum propaganda effect, rejecting Eisenhower's renewed proposal for Open Skies. Khrushchev declared "as long as arms exist our skies will remain closed and we will shoot down everything that is there without our consent." France's President Charles de Gaulle then asked whether this would include satellites, noting that Soviet satellites had already carried cameras into space. Khruschev replied, "As for sputniks, the U.S. has put up one that is photographing our country. We did not protest; let them take as many pictures as they want."[46]

The State Department continued to discuss the subject of freedom of space. By mid-1960, the Bureau of European Affairs at the State Department had drafted a policy paper concerning the planned upcoming launch of the SAMOS reconnaissance satellite. One option proposed was an open approach, which advocated the sharing of all reconnaissance photographs taken by SAMOS. This option was considered more likely to facilitate wide acceptance of photographic satellites than the closed approach.[47] While the State Department debated the merits of sharing reconnaissance satellite photographs with the world community (something that the United States had not done even with U-2 photographs), CORONA's engineers were preparing for another launch. The memo was written August 12. Six days later CORONA returned its first images of the Soviet Union.

The State Department continued to examine the benefits of the American space program, but gradually even SAMOS itself was enveloped within the dark cloak of secrecy that surrounded CORONA. State Department discussions of satellite reconnaissance in general, and public statements in particular, ceased. Freedom of space had been achieved. Further discussion was largely irrelevant. And CORONA began to chalk up success after success, as the press focused on SAMOS and the public paid attention to NASA and its daring Mercury astronauts. Early American space policy involved wrapping a riddle (CORONA) in a mystery, and surrounding it with an enigma.

GERALD HAINES

6
THE NATIONAL RECONNAISSANCE OFFICE
Its Origins, Creation, and Early Years

Formally established on September 6, 1961, as a super-secret, covert agency, the National Reconnaissance Office (NRO) has since developed and managed the revolutionary U.S. satellite reconnaissance effort.[1] Focusing its efforts primarily on the Soviet Union, for nearly 30 years the NRO provided U.S. policymakers and military planners with unique and essential raw intelligence on Soviet war-making capabilities that threatened or might threaten U.S. national security interests. The intelligence product from NRO reconnaissance satellite systems tracked Soviet weapon and missile developments, military operations, order of battle information, nuclear capabilities, and industrial and agricultural production.

It is difficult to exaggerate the value of this unique source of intelligence to U.S. policymakers and military planners.[2] The role played by the NRO during the Cold War was absolutely crucial. Moreover, the NRO and the satellite reconnaissance systems it developed radically changed the entire concept of intelligence. The satellite systems allowed the United States to collect an ever-increasing volume of detailed intelligence never before available and permitted U.S. civilian and military decisionmakers far more flexibility in reacting to potential Soviet threats during the Cold War. In addition, the NRO's unique mission drove innovative scientific and technological developments in space and created a model for government-industry cooperation in a crisis atmosphere.

This brief study outlines the origins of the U.S. reconnaissance effort in the 1950s and traces the creation and development of the NRO in the 1960s. It also examines in some detail the struggle between the Central Intelligence Agency (CIA) and the U.S. Air Force for control of overhead reconnaissance and the NRO. In 1991, the Cold War ended as the Soviet Union collapsed. In 1992, the Department of Defense declassified the fact of the existence of the NRO and thus officially recognized the existence of this unique organization. Only now may the vital role the NRO played during the Cold War be revealed. This is an initial part of that story.

ORIGINS OF THE NRO: THE U-2

The origins of the National Reconnaissance Office date back to the first cooperative efforts between the CIA and the U.S. Air Force to develop overhead reconnaissance systems. First was the U-2 reconnaissance aircraft, begun in late 1954 and developed by a small team consisting of the CIA, the Air Force, and several defense contractors. To achieve maximum security, CIA official Richard Bissell and Air Force Brigadier General Osmond Ritland made the project self-sufficient. It had its own contract management, administrative, financial, logistics, communications, and security personnel. Funding was also kept separate from other CIA or Air Force projects. Bissell reported directly to the Director of Central Intelligence (DCI), Allen Dulles. Bissell's use of "unvouchered funds" simplified competitive bidding procedures and significantly sped up the procurement process.[3] It simplified security because such funds did not need to be reported.

Bissell and Ritland gave Lockheed "performance" specifications for the U-2 rather than using the standard Air Force practice of giving "technical" specifications. According to Clarence L. "Kelly" Johnson, Lockheed program manager for the U-2, this allowed Lockheed to focus on performance goals rather than individual specifications. It gave the contractor greater flexibility in designing and building the aircraft. The arrangement became a unique partnership between the Lockheed Corporation and the government. Such unique streamlined management and acquisition practices were employed throughout the U-2's development, and set a precedent for NRO's approach to reconnaissance satellites. Time and results mattered, not bureaucratic paperwork.[4]

President Eisenhower authorized the project on November 27, 1954, and less than 10 months later, on August 5, 1955, the U-2 made its maiden flight. From June 20, 1956, through May 1960, the U-2 made a total of 24 overflights of the Soviet Union. Thousands of feet of film poured into the CIA's small Photo-Intelligence Division.

The photographs obtained by the first U-2 flights provided a bonanza of data for U.S. intelligence agencies. In fact, a photograph of the Saratov-Engels airfield at Ramenskoye, southeast of Moscow, taken on July 5, 1956, put to rest the "bomber gap" debate. It showed fewer than three dozen of the new Soviet Myasishchev-4 (Bison) heavy bombers. At the time, the U.S. Air Force was claiming that nearly 100 of the Bisons were already deployed. The U-2 missions could find no additional Bisons at other major Soviet airfields. DCI Allen Dulles referred to this photograph in later years as the "million-dollar photo."[5]

By 1957 U-2 missions were providing U.S. intelligence analysts with a wealth of information about Soviet missile, technological, and scientific activities. Known as Project SOFT TOUCH, these flights ranged over such prime Soviet targets as the missile-test facilities at Tyuratam and Sary Shagan and the nuclear refining installations at Tomsk.

ORIGINS OF THE NRO: PROJECT CORONA

In late 1954, at the same time the CIA was developing the U-2 reconnaissance aircraft and beginning to exploit overhead photography, Air Force officials called for continuous surveillance of denied areas of the world to determine a potential enemy's war-making capability. This rekindled the Air Force's interest in satellite development. On April 2, 1956, the WS-117L System Program Office (known as an SPO) of the Air Force published the first complete U.S. development plan for a reconnaissance satellite. The plan proposed a full operational system by late 1963 at a cost of nearly $115 million. The Air Force made Lockheed the prime contractor for this multifaceted space-system concept. A major component of the plan was a satellite observation system.

The Air Force envisioned a 92-satellite program divided into seven phases. The first phase would be a direct readout reconnaissance satellite that would process film aboard the satellite and transmit the images to a ground station. It was not to become operational, however, until 1960. The final phase, a large signals intelligence satellite, would be operational by 1963. The original price tag for the entire project rapidly escalated to $600 million. Little came of these efforts, however, as the Department of Defense struggled to eliminate noncritical defense expenditures during the mid-1950s and the Eisenhower administration stressed a "space for peace" theme. In addition, many civilian and military leaders doubted the reliability of such advanced concepts.

The launching of Sputniks I and II in the fall of 1957, however, changed the prospects of the reconnaissance satellite program. Although the new President's Board of Consultants on Foreign Intelligence Activities (PBCFIA), chaired by James Killian, assured the president that U.S. missile development was on track and on a par with or ahead of Soviet efforts, it urged greater federal support for the various programs and a major review of all reconnaissance systems with a view toward replacing the increasingly vulnerable U-2.

When Eisenhower asked Deputy Secretary of Defense Donald Quarles about the prospects for a U.S. satellite reconnaissance vehicle that could take pictures from space and beam them back to Earth, Quarles replied that the Air Force had a major research program in the area that was progressing nicely. Killian and another PBCFIA member, Edwin "Din" Land, disagreed. Killian considered the satellite program peripheral. He believed that if Project RAINBOW, designed to make the U-2 invisible to radar, proved successful, it would diminish the importance of satellites altogether.[6] Moreover, Killian was supported by James Baker, a Harvard University astronomer, head of the Air Force Intelligence Panel and, like Killian and Land, a member of the 1955 Technological Capabilities Panel that had recommended the development of the U-2. They were also joined by Philip G. Strong from the CIA's Office of Scientific Intelligence. These men believed that the major part of the Air Force's WS-117L project, the direct readout satellite, could not return the scale of imagery needed to answer the president's questions concerning Soviet missile developments.

Despite such concerns, Killian, Land, Baker, and their colleagues believed that U.S. scientists and engineers, given sufficient funds and the freedom to innovate, could solve the problems and get a film-return photoreconnaissance satellite in orbit. They also urged the president to start work on a replacement for the U-2 as soon as possible.

After listening to his advisers, on October 28, 1957, Eisenhower ordered the Air Force and the CIA to provide him with details of their efforts to date concerning advanced reconnaissance systems. For the CIA, this meant the supersonic reconnaissance aircraft, the OXCART or A-12 (the Air Force version was known as the SR-71 or Blackbird); for the Air Force, it meant the various satellites of the WS-117L project.

Discussions among the president, his civilian advisers, and CIA and Air Force officials continued into December. All agreed that there was little prospect that either the A-12 or WS-117L could be deployed soon. Nevertheless, the scientists, led by Killian and Land, urged the president to pursue both an advanced aircraft and the satellite projects. This investment in competing reconnaissance platforms corresponded with Eisenhower's belief that the nation would have to use a "Manhattan Project" approach in order to make rapid progress in the missile and satellite areas. Killian and Land also believed that a small element of the WS-117L program, a satellite with a returnable film capsule, could be quickly developed. They recommended that this program be taken from the larger Air Force project and given to the same team that had built the U-2: the CIA's Richard Bissell and the Air Force's Brigadier General Osmond Ritland. The civilian scientists believed such a move would take the pressure off the larger Air Force effort and serve as an interim reconnaissance system until the problems of the WS-117L could be worked out.

Under the covert plan approved by Eisenhower, the CIA would procure the satellite cameras and reentry vehicles, while the Air Force provided the host spacecraft and the booster missiles. At the same time, the CIA retained responsibility for developing a follow-on plane for the U-2 with the assistance of the Air Force.

The satellite program, Project CORONA, was to be a stop-gap effort until the much larger and more complex Air Force WS-117L was developed and deployed its satellites. Little did anyone realize the extent of the problems U.S. scientists would encounter in both programs, or that CORONA would become the pioneering program for unmanned space flight, or that it would still be launching satellites 14 years later.

The CORONA experience, like the U-2 program, also demonstrated to later NRO officials the advantages of a flexible government-business partnership arrangement. It essentially established the fundamental management and acquisition principles that the NRO followed for the next 20 years. Bissell later described the process:

> The program was started in a marvelously informal manner. Ritland and I worked out the division of labor between the two organizations as we went along. Decisions were made jointly. There were so few people involved and their relations were so close that decisions could be and were made quickly and cleanly. We did not have problems of having to make compromises or of endless delays awaiting agreement. After we got fully organized and the contracts had been let, we began a system of management through monthly suppliers meetings—as we had done with the U-2. Ritland and I sat at the end of the table, and I acted as chairman. The group included two or three people from each of the suppliers. We heard reports of progress and ventilated problems, especially those involving interfaces among contractors. The program was handled in an extraordinarily cooperative manner between the Air Force and the CIA. Almost all of the people involved on the Government side were more interested in getting the job done than in claiming credit or gaining control.

The CORONA program, like the U-2, used a tight-knit government-industry team approach. It provided maximum latitude to the engineers to grapple with technical problems and issues.

THE CREATION OF THE NATIONAL RECONNAISSANCE OFFICE (NRO)

Prior to the establishment of the NRO in 1961, the CORONA program operated under a loose, unstructured arrangement by which the CIA and the Air Force jointly ran the effort. The CIA handled the funds for the covert projects, acquired the CORONA cameras and the satellite recovery vehicles (SRVs), and provided much of the program's security procedures. The Air Force built the spacecraft, launched the rockets, and retrieved the payloads. For a time, the relationship worked well. CIA's Deputy Director for Plans (DDP) Richard Bissell and Undersecretary of the Air Force Joseph Charyk formed a close, high-level, informal working partnership.

According to Bissell, the program, despite some shortcomings, worked reasonably well. "Although I didn't like the situation," he remarked later, "I was perfectly well aware that it could not be any other way. The Agency could not get into the business of launching large missiles at that stage of the game."[7]

By the early 1960s, however, many in the Air Force concluded that their role should be preeminent, and tried to assert control over the entire project. From the Air Force's point of view, the service was doing 90 percent of the work. Hundreds of Air Force personnel were working on CORONA, while the CIA had only two officers stationed in California on the program—both Air Force lieutenant colonels on loan to the agency. For its part, the CIA wanted to maintain an independent capability for the design and development of satellite systems. Much of the early 1960s saw a struggle for control of U.S. reconnaissance satellites between CIA and the Air Force and attempts to work out some type of compromise agreement regarding the NRO.

PROBLEMS BEGIN

The conflict first surfaced over the introduction of the CORONA KH-3 (C‴) camera in mid-1961. When President Kennedy's science adviser, Jerome Wiesner, expressed reservations about the advanced camera, Colonel Lee Battle, the Air Force officer in charge of the launch facilities in California, arbitrarily canceled the first KH-3 launch scheduled for July 1961. Bissell's special assistant for technical analysis, Eugene P. Kiefer, criticized the cancellation. For Kiefer, the Air Force was intervening in areas in which it had no authority. He appealed to Bissell to have the launch reinstated. Bissell turned to his friend Charyk, who succeeded in rescheduling the KH-3 launch. The incident, however, caused both the CIA and the Air Force to rethink the management structure for the CORONA program.

Upset by the carping and complaining over the cancellation of the KH-3 launch, Killian and Land (who still served as intelligence advisors despite the change in administrations) suggested to DCI Allen Dulles and Deputy Secretary of Defense Roswell L. Gilpatric that the lines of responsibility in the CORONA program needed straightening.[8]

Neither Bissell nor Charyk was enthusiastic about signing a formal agreement. They believed that their informal collaboration over the past five years had provided them with needed flexibility and had avoided excessive bureaucracy. Nevertheless, Charyk asked an aide to draft a formal agreement. Unfortunately,

Joseph Charyk, the first director of the National Reconnaissance Office. (Photo courtesy of the National Reconnaissance Office)

the aide, Colonel John Martin, had never been actively involved in the CORONA program and knew little of the informal working relationship between Bissell and Charyk. His draft agreement carefully described the existing relationship in great detail, but still failed to capture the sense of informal cooperation that had made the relationship work. Bissell and Charyk both read the draft and made few changes. Both assumed they would continue to operate under their informal agreements.

After some discussion between Deputy Director of Central Intelligence (DDCI) Charles P. Cabell and Gilpatric, on September 6, 1961, the CIA and the Air Force officially signed a charter establishing a National Reconnaissance Program (NRP). Under the agreement, a covert National Reconnaissance Office (NRO) would finance and control all overhead reconnaissance projects. The NRO was to be managed by a joint directorship of the CIA and the Air Force reporting to the Secretary of Defense. It would obtain intelligence requirements through the United States Intelligence Board (USIB).[9] Budget appropriations for the central administrative office of NRO, made up of a small number of CIA, Air Force, and Navy personnel, came through the Air Force. Furthermore, the Air Force provided the missiles, bases, and recovery capability for the reconnaissance systems. The CIA, in turn, conducted research and development, contracting, and security. The agreement also left the CIA in control of the data collection program.

What the agreement did not address were the fundamental disagreements between the CIA and the Air Force over the entire satellite reconnaissance program and the very different objectives each had for this program. The Air Force, especially, was unwilling to relinquish control of what it viewed as one of its primary missions. By the 1960s, the reconnaissance program had assumed major importance for the Air Force. With the advent of intercontinental ballistic missiles, the manned bomber had lost its primacy in strategic planning. In addition, with the creation of the civilian National Aeronautics and Space Administration (NASA) space program in 1958, the Air Force lost its potential direction of the overall U.S. space effort. The Air Force was, therefore, reluctant to see overhead reconnaissance permanently snatched away as well.

Further complicating the situation was the very nature of the reconnaissance program itself. The Air Force was more interested in tactical intelligence, while the CIA paid more attention to procuring strategic or national intelligence. Also at issue were questions over requirements, who determined targets, and the frequency of coverage. If the Air Force assumed major responsibility, its decisions would reflect its tactical orientation; if the CIA decided, however, national intelligence requirements would have precedence.[10]

THE CIA–AIR FORCE STRUGGLE FOR CONTROL

The shake-up in CIA management which followed the failed Bay of Pigs invasion in 1961 made the situation even worse. By February 1962, Dulles, Cabell, and Bissell had all resigned or retired. The new DCI, John McCone, convinced of the importance of technical collection programs and under some pressure from President Kennedy's new Foreign Intelligence Advisory Board (PFIAB, which had replaced the PBCFIA, but was still headed by Killian) to consolidate management of CIA's technological development efforts, created a new Directorate for Research in 1962 and appointed Dr. Herbert "Pete" Scoville Jr. as its director.[11]

Added to the internal CIA turmoil over reorganization was the fact that the NRP agreement was quickly coming apart. With the departure of Bissell, the CIA had no real representation at the NRO meetings. Scoville, unaware of the close personal involvement of Bissell with Charyk in the overhead reconnaissance arena, detached himself from the effort. Instead he sent his deputy, Colonel Stanley W. Beerli, and, after his departure, Colonel Jack Ledford, to represent the agency at NRO meetings. They had little authority to act and were no match for the Undersecretary of the Air Force. Both, in fact, were junior to Colonel John Martin, now Chief of Staff of the new Air Force Space Systems Division.[12]

The cooperation that had so exemplified the U-2, OXCART, and early CORONA efforts vanished. Charyk soon complained to Gilpatric and McCone about the lack of cooperation. He believed CIA people were relying more on what the NRP agreement said than on getting the job done. Getting information on overhead reconnaissance from CIA officials was, Charyk told Gilpatric, "like pulling teeth." In discussions with Gilpatric and McCone in September 1961, Charyk indicated that the NRP agreement, signed only a few weeks before, needed clarification.

In May 1962, just 10 weeks after Bissell's departure, McCone and Gilpatric signed a second NRP agreement that enumerated more clearly the responsibilities of the NRO for conducting the National Reconnaissance Program. The new agreement gave the NRO control over all reconnaissance spending, including funds that were part of the CIA's budget. The agreement also established a single director of the NRO who would be jointly appointed by the Secretary of Defense and the DCI. McCone readily accepted the concept of appointing Charyk the first director of the NRO (DNRO), but was not enthusiastic about his successors coming from the Defense Department. In return for accepting the Undersecretary of the Air Force as the single director of the NRO, McCone demanded assurances on continued CIA control over research and development, contracting, and targeting.

The new agreement made no mention of a deputy director, however. Charyk, convinced that a deputy would lead to yet another layer of bureaucracy and that there was not enough work for two people, opposed the creation of a deputy slot. Although McCone did not object to Charyk's position, the lack of a deputy soon caused additional problems and friction.

As expected, Secretary of Defense Robert McNamara and McCone named Charyk DNRO.[13] Charyk's first directive attempted to deal with the deep divisions within his organization by creating separate programs. He established a Program A (USAF satellite assets), Program B (CIA assets), Program C (U.S. Navy assets), and Program D (USAF aircraft assets).[14] While Charyk hoped this would stop the bickering, it did not.

The CIA saw its role in satellite reconnaissance eroding. Many in the CIA looked upon the NRO as a thinly disguised extension of the Air Force. DCI McCone was unwilling to concede all reconnaissance programs to the Air Force. In addition, he did not want defense requirements to overwhelm national intelligence requirements. For McCone, the NRO was a national asset, not simply a tool for the military. Urged on by Albert Wheelon, who had replaced Scoville as DDR, McCone began to challenge the Air Force and the NRO and to question their ability to run the satellite programs effectively.[15] He pointed out that the Air Force was responsible for most launch failures in the program to date and accused McNamara and Gilpatric of spending too much time defending the TFX fighter plane proposal before Congress rather than concerning themselves with the complex problems of overhead intelligence collection.

For its part, the Air Force now moved to secure control over the entire reconnaissance effort. For example, in 1962 Dr. Brockway McMillan, Charyk's successor as DNRO, supported the Air Force position when he recommended that the entire photoreconnaissance satellite program be turned over to the Air Force in order to streamline command relationships and achieve greater success. For McMillan, the NRO was primarily an Air Force activity and the CIA was being irrational and obstructionist when it came to working on satellite reconnaissance. The rivalry between the Air Force and the CIA intensified.[16]

Ironically, DNRO McMillan was not a strong supporter of the Air Force when it came to satellite reconnaissance capabilities. McMillan trusted neither the Air Force nor the CIA. Given the recent agreement, he believed the NRO should control the satellite programs. It was that simple.

The Air Force and DNRO McMillan's drive to completely control the reconnaissance program actually jeopardized the Secretary of Defense's capacity to utilize reconnaissance data. In order to make independent judgments on weapons procurement and strategic planning, Secretary McNamara decided he needed an independent analytical capability in the Office of the Secretary of Defense. If the Air Force controlled the reconnaissance program completely, it would have an enormous advantage in pressing its own claims. McNamara, aware of the threat, often sided with McCone against the Air Force in order to maintain his own position as arbiter of DoD planning and resource allocations.

This led to a third NRP agreement, signed by McCone and Gilpatric in March 1963. This time the duties and responsibilities of a deputy director of the NRO were carefully spelled out, with the expectation that a CIA officer would fill the slot. Wheelon appointed Eugene Kiefer to the position. The agreement also stated that the Secretary of Defense was to be the executive agent for the NRP and that the NRO was under the direction, authority, and control of the Secretary of Defense. The NRO was to be developed, managed, and conducted jointly by the Secretary of Defense and the DCI and it was to be a highly secret, separate operating agency of the Department of Defense, no longer under the control of the Office of the Secretary of the Air Force.

Despite the new agreement, the Air Force continued to press for complete control of the overhead reconnaissance programs. McMillan was caught in the middle. He wanted a strong NRO and often clashed with both the Air Force and the CIA. McMillan's determination to make the NRO the leading organi-

Dr. Albert "Bud" Wheelon, Deputy Director of Science and Technology at the Central Intelligence Agency from 1963 to 1966. Wheelon fought for continued CIA involvement in satellite reconnaissance in the mid-1960s and built the Directorate of Science and Technology into a powerful engineering development center. (Photo courtesy of Albert D. Wheelon)

zation in satellite development and the antagonism that had grown between the CIA and the Air Force hampered the NRO decisionmaking process during 1964. Conflicts arose over contracting, funding, and the CIA's role and responsibilities in the reconnaissance area. Added to the problem was a major personality conflict between Wheelon and McMillan.[17]

The situation deteriorated so much that DCI McCone and new Deputy Defense Secretary Cyrus Vance finally agreed to meet as an NRP executive committee in order to make funding decisions for the NRP. McCone suggested to McNamara at the same time that the only way to solve the problem was to remove the NRO completely from the parochialism of the Office of the Secretary of the Air Force and place it firmly in the Office of the Secretary of Defense. McCone constantly complained to Dr. Eugene Fubini, director of DOD's Defense Research and Engineering, that he "never knew the first damn thing that was going on with regard to the NRO budget." McCone feared that the entire overhead reconnaissance program was becoming little more than an instrument of the Air Force and that national intelligence requirements were sinking to second, third, or fourth priority. McMillan also complained to Fubini that the CIA was attempting to undermine the NRO by refusing to disclose program data.

MORE PEACE TREATIES

Before any actions were taken, in April 1965 the formidable McCone resigned as DCI and President Lyndon Johnson replaced him with Vice Admiral William F. Raborn Jr. The CIA's battle with the Air Force and McMillan was the first major issue confronting the new DCI. Raborn appointed John Bross, the director for the National Intelligence Programs Evaluation, to negotiate a settlement along the lines suggested by McCone.[18] Bross's efforts resulted in Raborn and Vance signing a fourth NRP agreement in August 1965, which recognized the need for a single national satellite reconnaissance program to meet the intelligence needs of the United States. It gave the DCI and the Secretary of Defense decisionmaking authority over all national reconnaissance programs. It established the NRO as a separate agency within the DoD and officially designated the Secretary of Defense as the executive agent for the NRP. It also set up a three-person executive committee (EXCOM) for the management of the NRP. The EXCOM membership included the DCI, the Deputy Secretary of Defense, and the president's science advisor. The EXCOM reported to the Secretary of Defense. The new arrangement also recognized the DCI's right as head of the Intelligence Community to establish collection requirements in consultation with the USIB.[19]

The agreement represented a compromise between the Air Force and the CIA. It led to the CIA and the Air Force cooperating successfully on several satellite collection projects. As a decisionmaking structure, it worked well. The compromise agreement, however, left the inherent competition between the two organizations over satellite collection systems intact, a situation not entirely detrimental to the development of the U.S. satellite program. Urged on by their rivalry and a sense of national mission, both the CIA and the Air Force pushed the cutting edge of technology in satellite development and data return from space.

INNOVATIONS AND SUCCESSES

The Soviet threat and the Cold War dominated U.S. foreign policy for nearly half a century. For much of that time, the NRO and its revolutionary overhead reconnaissance systems, whose very existence remained classified, were the single most important source of information on Soviet military programs and capabilities. The NRO produced, according to some estimates, nearly 90 percent of all intelligence data on the Soviet Union during this period. NRO satellite systems established, with considerable accuracy, the actual military capability and preparedness of the Soviet Union.[20]

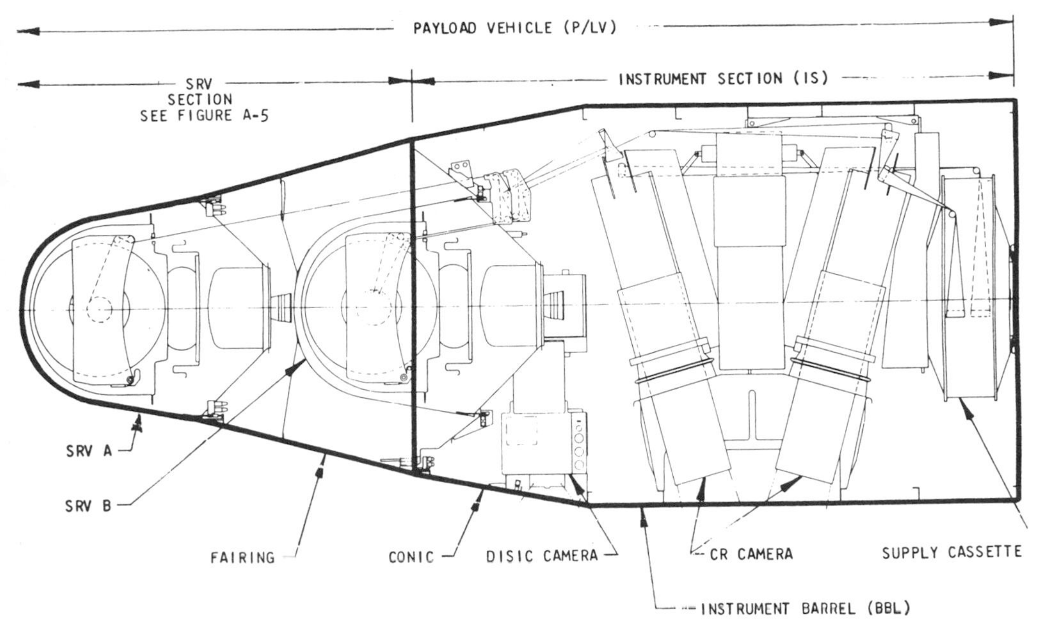

The KH-4B was the most advanced version of the CORONA satellite. Representing the limits of the CORONA design, it operated from 1969 to 1972.

Costs were rarely questioned. The NRO mission held the highest national priority. The NRO was to gather intelligence on "denied areas," especially the Soviet Union, which represented the greatest threat to U.S. national security. This mission engaged not only the top political leaders, including the president, but also the nation's major intelligence and military officials, as well as its senior scientists and defense industrial talent. The NRO and its satellite reconnaissance systems radically changed the entire concept of intelligence gathering. It allowed the United States to collect an ever-increasing volume of detailed intelligence vital to U.S. national security interests.[21] There is little doubt that the NRO played a major role in the U.S. victory in the Cold War.

CONCLUSION

Despite its problems, the NRO worked. Its mission was of preeminent importance to the United States. Nevertheless, for some years after its origin in 1961 the NRO experienced constant turmoil and crisis. Organizationally, its most important feature during the Cold War years was that it consisted of three different program offices that were managed by three different agencies: the CIA, the Air Force, and the Navy (Program D was eliminated in the late 1960s). This produced natural and, at times, heated competition. CIA/Air Force elements of the infant NRO fought constantly for control over overhead reconnaissance systems.

The result of a "forced marriage," the NRO, nevertheless, operated in a national interest that transcended the parochial views of the CIA, Air Force, and Defense Department. It operated continuously in a crisis atmosphere. Despite its enormous successes, however, it was never able to resolve the fundamental intelligence differences represented by the CIA and the Air Force regarding a strategic or tactical intelligence focus. Efforts by the Department of Defense and the U.S. military command structure to gain greater control over space reconnaissance development and operations continued.

Nevertheless, with its critical national security requirements, the NRO developed unique, flexible, and streamlined acquisition and management procedures to build, operate, and maintain a U.S. space reconnaissance capability. Streamlined procurement practices used by the NRO cut approximately one year out of the time it would otherwise have required to bring a satellite system on-line. The NRO brought the best engineering and designing talent in government and private industry together in a unique partnership. The U.S. aerospace industry actually performed much of the research and development, design engineering, systems integration, manufacturing, testing, launch integration, and much of the actual operation of the NRO's satellite systems under a flexible government-industry partnership.

In the early years of CORONA, the NRO and its private industry partners created the technology needed to produce imagery and intercept signals from space. The NRO pushed photoreconnaissance technology from fuzzy, barely identifiable images with low resolution to clear, detailed photographs with high resolution. In addition, the NRO went from reconnaissance satellites that it hoped would last a few revolutions around the earth to satellites that remain in orbit for years. It set the stage in communications technology for the information superhighway revolution. Operating during the crisis atmosphere of the Cold War, the NRO and its reconnaissance satellites were truly a remarkable intelligence achievement. They allowed U.S. policymakers and military planners to keep a close, accurate watch on the Soviet Union throughout much of the Cold War.

PETER A. GORIN

7
ZENIT
The Soviet Response to CORONA

The space program in the Soviet Union developed much along the same lines as in the United States. Both countries knew from the start that an orbital perspective would offer unparalleled opportunities for gathering information about activities on the earth's surface. As in the United States, strategic reconnaissance was an early objective of Soviet space efforts.

A LONG ROAD TO SPACE

The first Soviet study of artificial satellites was initiated by a group of engineers under Colonel Mikhail Tikhonravov at the military rocket research institute, NII-4, from 1948 through 1950.[1] Not officially sponsored, the study nearly cost Tikhonravov his career when he gave his proposals to the Soviet high command. In 1950, he was demoted from deputy director of NII-4 to scientific consultant.[2] At that time, most political and military officials considered satellites a no-value deviation from the primary task of the Soviet rocket industry, which was the development of ballistic missiles.

Dr. Tikhonravov had long been dedicated to the idea of rocket flight into space; in 1933, he had designed and launched the first Soviet liquid-propellant rocket. After his demotion in 1950, he continued his satellite research with a small group of enthusiasts and lobbied the government with proposals for rocket flight. A number of high-ranking officials from the defense industry supported their efforts, especially Mstislav Keldysh (director of NII-1, research institute of Ministry of Aviation Industry) and Sergei Korolev (Chief Designer of the Experimental Design Bureau [OKB-1] of the Ministry of Defense Industry).

Korolev and Tihkonravov had known each other since the early 1930s and both had worked on rocket research before World War II.

Keldysh and Korolev received sufficient support from the Soviet Academy of Science for their idea of an artificial satellite. On May 27, 1954, Korolev wrote to Minister of Defense Industry Dmitriy Ustinov proposing that a satellite be launched by a future ICBM. The letter was supplemented with a detailed proposal for various types of satellites, which was prepared by Tikhonravov with Korolev's and Keldysh's input.[3] Just a week earlier, the government had officially adopted a program for the first Soviet ICBM, R-7 (U.S. code name: SS-6), to be directed by Korolev. As a result of this proposal, the government in August 1954 officially began to sponsor preliminary satellite research.[4]

Korolev and some in the Soviet leadership apparently realized the propagandistic importance of being the first in space during the International Geophysical Year (IGY). (The USSR openly proclaimed that intention in 1956, but the Western world apparently did not pay attention.)[5] The launch of the satellite was intended to demonstrate the capability of the ICBM even before it became operational, an idea consistent with the Soviet strategy of "psychological deterrence" of the anticipated American nuclear attack. The idea of using a satellite for reconnaissance was probably proposed during the period of preliminary satellite research in 1954. The feasibility of artificial satellites indicated that their military applications would become feasible as well.

The artificial satellite development program was adopted by a secret government decree on January 30, 1956, authorizing the development of a scientific, uncontrolled satellite, designated Object D. That project resulted in the launch of Sputnik-3 on May 15, 1958. The first two Sputniks, designated PS and PS-2, were simplified spinoff projects of Object D, designed solely to beat the Americans to space. Tikhonravov, one of the pioneers of rocket flight, was initially a consultant for the satellite program, but in early 1957 he was put in charge of all space projects of OKB-1 as the head of the newly established Department-9.[6] Consequently he became the "father" of the world's first artificial satellite.

Work on the photoreconnaissance satellite was apparently authorized by the same decree. Judging by Sergei Korolev's letter to the government of April 12, 1957, substantial preliminary research had already been done on another project, Object OD-1, a three-axis stabilized satellite for photo-imaging from space—an obvious photoreconnaissance prototype.[7] OD-1 was similar to its American counterpart, CORONA. It had a large, roughly cylindrical body containing a camera with a 1,000mm focal length and all the support systems in it. The conical-shaped return capsule was to carry the exposed film and the minimal number of systems essential for reentry and recovery.[8] The OD-1 orbital mass

The TSSKB/Progress Factory, Samara, Soviet Union. This CORONA image depicts the site where the Soviets manufactured their Zenit reconnaissance satellites. (Photo courtesy of C. P. Vick)

was limited to 1,500 kg. Heavier variants (for another launch vehicle) were considered as well.

Conditions for the project's implementation were not ideal. The reconnaissance satellite's development paralleled several other top-priority programs at OKB-1. Department-9 was involved in preparing the first Sputniks and early lunar probes, which were in high demand for propaganda. It was determined in 1957 that, because the reconnaissance satellite weighed 400 kg more than the rocket could lift, the launch vehicle would need to be modified for use with OD-1.[9] A modified launch vehicle was developed, but was never used for OD-1 because of radical changes in the satellite's design. The only space mission for this modified booster was to launch Sputnik-3. As a result of the higher priority of other projects and the sheer complexity of the photo-imaging satellite, between 1956 and 1960 OD-1 progressed only as a paper study.[10]

In late 1958, Korolev initiated a complete redesign of the satellite, making it compatible with another ambitious project of his design bureau—Object OD-2. The OD-2 project, which started in early 1958, was a study of a heavy three-axis stabilized satellite for human spaceflight. The rationale behind this decision was simple: the OD-1 and the OD-2 projects had one common feature—they were supposed to ensure the safe reentry and return of the capsules. Why develop two different systems for the same purpose? The exposed film could be returned by the same type of reentry vehicle that would carry a cosmonaut.

ZENIT-2

Soviet government decrees on May 22 and 25, 1959, authorized the parallel development of a manned spacecraft (3K), its simplified prototype (1K), and the photoreconnaissance satellite (2K); all were based on the same design, the OD-2 project.[11] All these spacecraft received the additional designation Vostok (Russian for "East"), and were respectively called Vostok-3, Vostok-1, and Vostok-2. The name Zenit-2 (Zenith) was apparently first applied to the Vostok-2 (2K) reconnaissance satellite in 1961, when the name Vostok became publicly known as the world's first manned spacecraft. Zenits received the designation number 2, and later the number 4, because their original projects were called 2K and 4K. Hence there was no Zenit-1 or Zenit-3.

OKB-1, now known as Rocket-Space Corporation Energiya (RKK Energiya), was the primary contractor for the photoreconnaissance satellite and its launch vehicle, while the Ministry of Defense was OKB-1's primary customer. The technical requirements for the spacecraft and its mission profile were prepared by the NII-4 military research institute under Lieutenant General Andrey Sokolov. Major mathematical support, especially in ballistics, was provided by the Department of Applied Mathematics (headed, along with NII-1, by Dr. Mstislav Keldysh) from the Steklov Mathematics Institute of the USSR Academy of Science. Other institutions of the Academy of Science did additional theoretical research as well. The Institute of Applied Geophysics conducted a study on locating and recovering returned capsules. Vavilov State Optical Institute studied suitable optical systems.[12] In all, more than 120 various scientific institutes and production plants from the Soviet military and civilian industry participated in the photoreconnaissance satellite program.[13]

Zenit and the manned Vostok shared reentry and service modules as well as a launch vehicle, but the two spacecraft were not identical. Only nine out of Zenit's nineteen primary systems were borrowed from Vostok.[14] The rest had to be developed from scratch. In 1961, a group of satellite designers was separated from Department-9 and reorganized into the specialized Department-29 under Evgeniy Ryazanov. Zenit-2 was developed at that department under Yuri Frumkin (the section chief) and Boris Rublev (the leading designer). Pavel Tsybin, Korolev's deputy, supervised the project.[15] No novice in strategic reconnaissance, Tsybin had directed an ambitious but unsuccessful supersonic strategic reconnaissance aircraft project (RSR) in the late 1950s, prior to his transfer to OKB-1.

Zenit was the most complex Soviet space project of the time, exceeding even the manned Vostok. Many particularly difficult technical problems had to be resolved. Zenit required a sophisticated three-axis stabilization system, which was required for pinpointing the cameras. Unlike the manned Vostok, Zenit-2 required constant orientation at all times. New gyroscopic stabilizing platforms and infrared sensors had to be developed for the spacecraft. The orientation control system included an array of cold gas thrusters with a large load of compressed nitrogen, which was stored in a collar of spherical tanks between the return and service modules. A group of engineers under Boris Rauschenbach and Evgeniy Tokar from the NII-1 scientific research institute pioneered the development of space orientation systems in the Soviet Union.[16] The first of these systems was used aboard the Luna-3 spacecraft, which took pictures of the moon's dark side in October 1959. In 1960, Rauschenbach's group was transferred to OKB-1 and developed a more complex orientation system for Zenit.

Unlike their American counterparts, Soviet engineers traditionally preferred to shield the most sensitive systems of their satellites from the space vacuum. Creating an artificial atmosphere inside pressurized containers was easier than designing new vacuum-rated equipment. While ensuring higher reliability, this approach had a price, since the resulting satellites were larger and heavier. The Zenit's cameras worked in an artificial environment inside the return module. Temperature change substantially influenced image quality; the environmental control system was required to keep air temperature fluctuations to a maximum of no more than 0.1 degrees Celsius per hour.[17] Since the optical transparency of the thick porthole glass was affected by internal pressure and external contrasting temperatures, the cameras had special adjustable lenses. Movement of the satellite also created a problem of image displacement—especially for lenses with long focal lengths. Image motion compensation was ensured through a slow motion of the film itself, which was adjustable to the rate of terrain motion in the camera's view.[18] The necessity for constant control and adjustment of the satellite's equipment required a new multichannel telemetry and radio-command system with a capability to encrypt information. Radio exchange between Zenit and a ground control center was expected to be 10 times more intensive than that of the manned missions.[19]

The Zenit-2 final layout was very different from CORONA. The Soviet satellite was heavier (up to 4,740 kg) and was separate from the third stage of the launch vehicle. Its two modules were about 5 meters long and 2.43 meters in diameter. The nonpressurized instrument module housed electrical batteries, equipment for orientation, environmental control, and radio communications, and a liquid-propellant braking engine. That engine, called TDU-1, was not restartable, and the satellite was incapable of changing orbits. The instrument module was similar to that of Vostok, but with the addition of a cylindrical adapter between two conical shapes. Thus, its increased length was about 3 meters (2.43 meters in diameter) and the module weighed about 2,300 kg. A large spherical return (reentry) module (2.3 meters in diameter and about 2,400 kg in weight) was pressurized and contained cameras, film, parachutes, and an emergency destructive device to prevent the module from being captured by a foreign country.

The Zenit-2 return module accommodated several different camera arrangements. Initially it was equipped with one camera, called SA-20, with a 1,000mm focal length lens and a second camera, SA-10, with a 200mm lens. It also had a special photo-television system, called Baikal (after a Siberian lake). The latter was a film readout device that scanned photo images and transmitted them to the control center electronically. A similar device had been used by the Luna-3 probe in 1959 to produce pictures of the moon's dark side. The film readout system was installed on six early test models of Zenit-2 in 1961–63. Four of those satellites (Kosmos-4, 7, 9, 12) tested that system in space (two were launch failures). The Baikal system was apparently an early attempt to obtain "near-real-time" image return, but its performance as a reconnaissance tool was not satisfactory. Hence, the later models of Zenit-2 relied on film cameras only.[20]

The operational camera arrangement for Zenit-2 was called Ftor-2R ("Fluorite"). It consisted of four fixed cameras: three 1,000mm SA-20 units and one SA-10 with a 200mm lens. The latter camera was used for low-resolution pictures—to provide a location reference for the high-resolution images taken at the same time. Removal of the readout system allowed the designers to increase the film load of each camera to 1,500 frames. Unlike CORONA's panoramic images on narrow film strips, the SA-20 made pictures on square frames (apparently 300 × 300 mm). That made possible stereoscopic flat field photographs. The image format of the smaller camera probably was 180 × 180 mm.[21] With an average flight altitude of 200 km, the main cameras scanned a terrain strip 60 × 180 km. Hence, each camera covered an area of 3,600 square kilometers (60 × 60 km) per frame. Dr. Frumkin, one of the Zenit-2 principal designers, stated that an area of about 10 million square kilometers (more than the entire territory of the United States) could be photographed in each mission.[22] Various Russian publications provide conflicting data on the camera's ground resolution. Officially it was

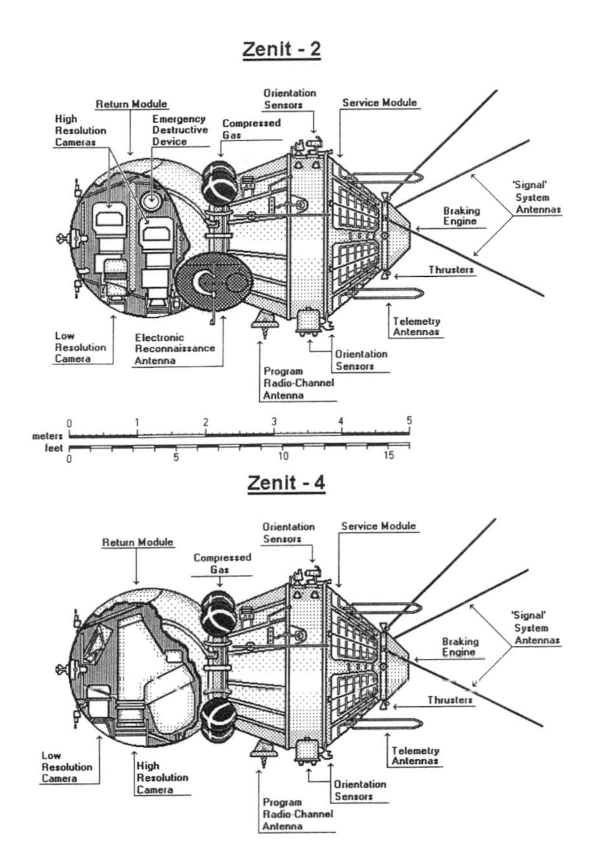

The Soviet Zenit photoreconnaissance spacecraft. Unlike its American counterpart, Zenit returned both cameras and film to Earth. (Illustration courtesy of Peter Gorin)

said to be about 10–15 meters, but some sources have suggested much better resolution.[23]

The Zenit cameras were developed by the Krasnogorsk Optical-Mechanical Plant near Moscow.[24] For decades that enterprise—one of the largest of that type in the USSR—was known for commercial production of the single-lens reflex cameras called Zenit. A freely available photo-camera was named after a super-secret military program!

In orbit, Zenit kept a horizontal orientation (its longitudinal axis was pointed along the flight path). Since the satellite flew in a low orbit, this orientation was useful to minimize drag from the atmosphere.[25] The cameras looked through the side portholes of the return module. Pictures were taken in three different modes: continuous scanning (with all three cameras constantly taking photographs), a single picture (three-frame strip) along a ground track, or a single picture on either side of the flight path. For the latter mode, the satellite performed a controlled roll, using its vernier thrusters.

Zenit satellites were normally launched to a near-circular orbit with an average altitude of 200 × 350 km. Orbital inclination was 51.8–65 degrees (from Tyura-tam) and 72–81 degrees (from Plesetsk), depending on the target location. Due to the natural drift of the orbit, the altitude of each mission was preselected so that the satellite would cover the same area twice in seven days.[26] The Zenit's layout made it impossible to have several return capsules, as was done on later versions of CORONA. Thus, a standard mission lasted for eight days, although it could be prolonged to up to twelve days. At the same time, Zenit's layout had a substantial economic advantage: the flown cameras were reused several times.[27]

Apart from overhead photography, Zenit-2 was designed to perform signals intelligence—interception of U.S. and NATO air defense radar frequencies. For that purpose, the satellite was equipped with a special system, Kust-12M ("Bush"). Intercepted information was recorded and stored aboard until the next communication session with the ground control center. Zenit-2 had a high-gain parabolic antenna, which was used for downloading electronic reconnaissance data.[28]

The USSR achieved its first successful space recovery on August 20, 1960—just nine days after the first American recovery of Discoverer XIII. That recovery of the Vostok prototype (1K-2), carrying two dogs, signaled a green light to both the manned Vostok and automated Zenit programs. Apparently, the Soviet leadership gave the manned program a higher priority, since it promised greater political advantages. However, in some of the launches of unmanned Vostok prototypes, experiments were conducted on photography from space.[29]

Zenit-2 had a rocky start: the launch vehicle failed on November 11, 1961, during the first Zenit-2 launch attempt. Contrary to Western reports, the spacecraft was not lost in Siberia but was destroyed by an emergency destructive device.[30] The Zenit-2 test program resumed on April 26, 1962, when the second satellite was put into orbit under the false designation of Kosmos-4. The orientation system malfunctioned and the cameras did not produce satisfactory pictures, but reentry was successful after a three-day flight. The first reconnaissance pictures from space were obtained by the USSR from the third Zenit-2 (Kosmos-7, July 28–August 8, 1962). This was almost two years after CORONA returned its first images.

In the first two launches of Zenit-2, the launch vehicle used was the same as for the manned Vostok. That was the three-stage version of the R-7 ICBM (U.S. code name: SL-3). Later, Zenit-2 satellites were launched by another booster, the three-stage version of the improved R-7A ICBM.[31] Twice (on June 1, 1962, and July 10, 1963) the rockets, along with their Zenit satellites, exploded on a launch pad at Tyura-tam, causing substantial damage to ground facilities and delays in test flights.

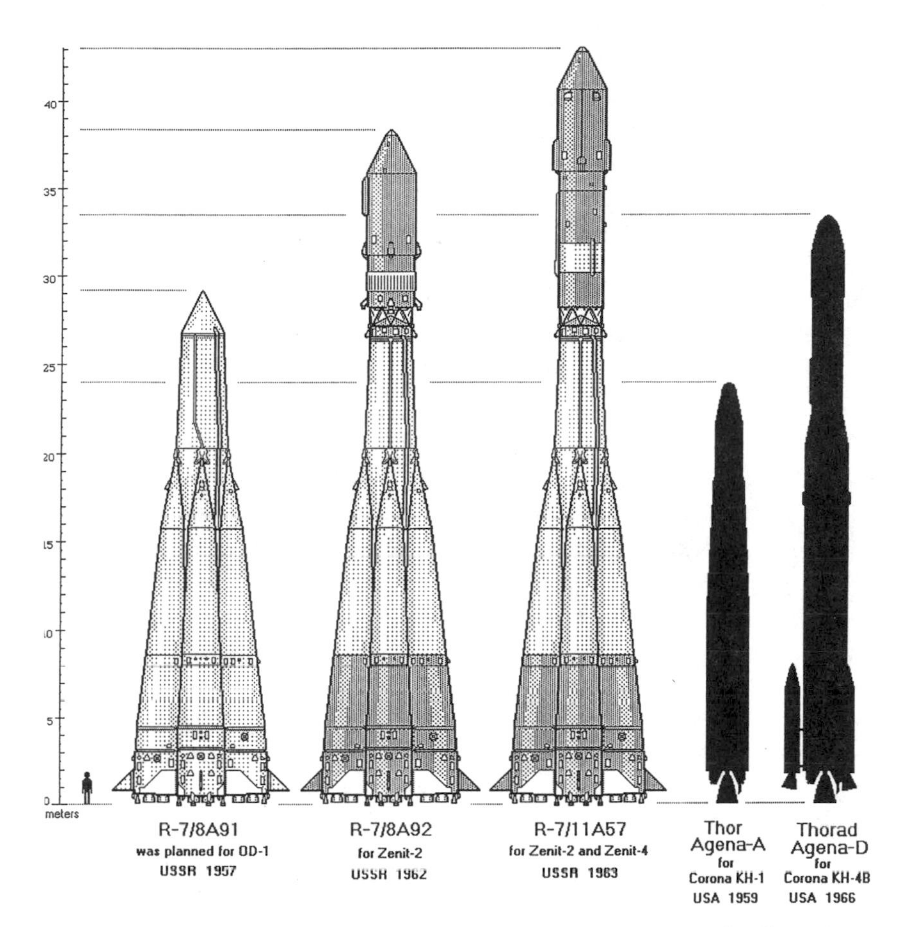

Launch vehicles for Soviet and American photoreconnaissance spacecraft. (Illustration courtesy of Peter Gorin)

Western analysts usually consider 1962 the starting point of Soviet reconnaissance activities in space. This is not quite correct. It took nearly two years (November 1961–October 1963) and thirteen satellites (three of them failed at launch) to complete the test program.[32] The last Zenit-2 test vehicle (Kosmos-20) was launched on October 18, 1963.[33]

SOVIET NATIONAL TECHNICAL MEANS OF VERIFICATION

That the United States was developing photoreconnaissance satellites was not a mystery to the Russians. In the early 1960s, Soviet propaganda viciously criticized the United States for "espionage from space." Prime Minister Nikita Khrushchev promised to shoot down American spy satellites. This was not just a bluff—the USSR was developing antisatellite weapons. The Soviet Union even sponsored a UN General Assembly resolution on banning espionage from space. As the Zenit-2 system became operational, however, Soviet antisatellite propaganda was gradually put to an end. For almost twenty years the governments of the Soviet Union and the United States pretended that space reconnaissance systems did not exist. This so-called secrecy was intended to fool not the potential enemy but the general public. In major arms reduction agreements both superpowers not only acknowledged but stressed the ultimate importance of what they called "national technical means" for treaty verification.

In the USSR, the system of national technical means of verification—space reconnaissance—was declared operational on March 10, 1964, when the Ministry of Defense commissioned the Zenit-2 complex by a secret decree.[34] Apart from the satellite itself, that complex included several other major elements.

Starting from Sputnik, all space missions in the USSR were prepared and launched by special military units—the "space troops"—of the Ministry of Defense. In December of 1959, those units were incorporated into the newly organized Strategic Rocket Forces (RVSN). The military also controlled all three rocket/space test ranges: Kapustin Yar (from 1947), Tyura-tam (from 1955), and Plesetsk (from 1957). Initially, Zenit missions were launched from Tyura-tam (better known as "Cosmodrome Baikonur"), where two launch pads for the R-7 rocket had been constructed. Starting from 1966, the spy missions more heavily relied on Plesetsk in the northern part of Russia. Plesetsk had four R-7 launch pads that had been used as operational R-7 ICBM sites until about 1965.

The reception and processing of data from all space objects was controlled by the military as well. From 1956, the NII-4 supervised the development of the ground-control station network. The most prominent role in that development was played by Colonel Yuri Mozzhorin, then–deputy commander of the NII-4. (Later, by then Lieutenant General Mozzhorin for thirty years headed TsNIIMASH [former NII-88]—the leading research center of the Soviet rocket/space industry.)[35] The Command and Data Processing Center (KIK) was established in the city of Krasnoznamensk (code named "Golitsino-2"), Moscow Region, while the reception and relay stations were placed along the whole territory of the USSR—from the Black Sea to the Pacific. Initially that system was used to track ballistic missile tests, but it was later adapted for space missions. The launch control center at Tyura-tam and KIK were among the first in the USSR to be equipped with first-generation Soviet-made computers. The KIK became fully operational in May 1958, when it processed data from Sputnik 3. The reception and data processing network was substantially modified and enlarged in 1959–60 to accommodate the Vostok and Zenit programs. In addition to new ground stations, satellite tracking capability was enhanced by introduction of four floating stations aboard modified commercial ships.[36]

To ensure the timely recovery of cosmonauts and evacuation from the Vostok landing sites, a special Search and Rescue Service was organized by the Soviet Air Force in 1960. The service included seven groups of paratroopers (3–4 persons each) and was equipped with twenty large passenger planes and ten helicopters.[37] The same Search and Rescue Service probably was also involved in recovery operations for Zenit which, unlike the CORONA reentry vehicles, came down on land.

Zenit-2 was the first military space system employed by the USSR Ministry of Defense on a regular basis. Growing military activity in space required greater coordination. Hence, the Central Directorate of Space Systems (TsUKOS) was established under the Strategic Rocket Forces in October 1964.[38] It was a precursor of the modern Military Space Forces of Russia. The TsUKOS incorporated the launch facilities, NII-4, KIK with its network of control stations, and the RVSN Directorate of Satellites and Space Objects. It also became the major procurer of new military space systems.

According to Frumkin, the results of the first Zenit-2 photographs exceeded even the most optimistic expectations.[39] The consumer of space reconnaissance data apparently was Soviet military intelligence—the GRU (Chief Intelligence Directorate) under the Soviet Armed Forces General Staff. Strategic reconnaissance in the interests of all services of the Soviet Armed Forces was the GRU's prime objective. Some sources indicate the existence of a GRU Satellite Intelligence Directorate, which interpreted and analyzed space photos, providing results to the high political and military leadership.[40] Other presumed photoreconnaissance consumers included the Topographical Directorate of the USSR Armed Forces General Staff and the Intelligence Department of the RVSN Commanding Staff. The first was responsible for military mapping; the latter needed satellite information for precision ICBM targeting.

To disguise military space missions, the Russians actually used an approach similar to the American Discoverer coverup, but they did so in a more elaborate way. They concealed not just one military project under a false cover, but all of them! They mixed various military and civilian satellites, failed probes, and other orbital objects, and called them all part of the Kosmos program. That program represented about 95 percent of Soviet space missions. At least one-third of all Kosmos satellites in reality were photoreconnaissance spacecraft.[41] Nearly all launches were announced, but it required a painstaking analysis of orbital and other parameters to distinguish one Kosmos satellite from another. Very few actual scientific or applications satellites received different designations.

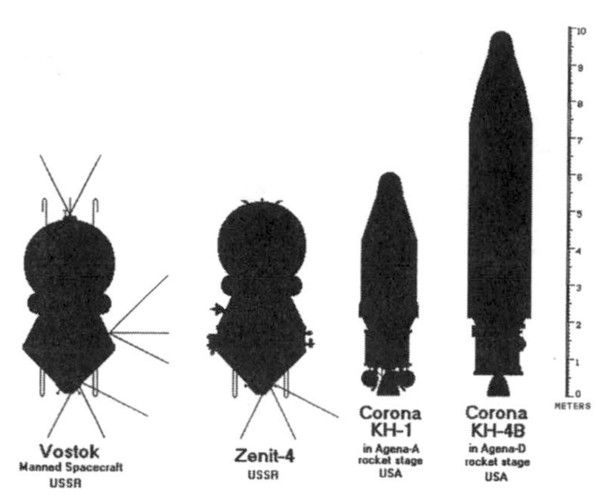

A comparison of the American CORONA and Soviet Zenit photoreconnaissance spacecraft. Some specific details of the Zenit remain unknown, but it was roughly comparable to CORONA and served the same role. (Illustration courtesy of Peter Gorin)

In 1968, a modified version—Zenit-2M—was introduced. There is no data available on the improvements to the new satellite. It is only known that the satellite incorporated a new launch vehicle and some improved systems from its sister spacecraft—Zenit-4 (discussed below). The signals intelligence equipment was apparently removed.

The Zenit-2 program remained operational for almost seven years and was retired after the launch of Kosmos-344 (May 12, 1970). Satellites of this type (including test vehicles and modified versions) were launched eighty-one times with fifty-eight completely successful missions. Eleven missions were only partially successful. Twelve launches failed, seven from launch vehicle failures and five from malfunctions of satellite systems.[42]

ZENIT-4

Even before Zenit-2 became operational, OKB-1 began development of a modified version, Zenit-4 (4K). Detailed technical data of Zenit-4 has not yet been revealed, except for a single cutaway drawing. Instead of three cameras with 1,000mm lenses, Zenit-4 apparently had one camera with a 3,000mm lens. In order to install such a large telescope into the reentry vehicle (which had a diameter of only 2,300 mm), the designers folded the camera's light path, bending it backwards.[43] It is also possible that this high-resolution variant was only one of several models of Zenit-4.

Although it had the same dimensions as its predecessor, Zenit-4 was much heavier—at least 5,500 kg (compared to 4,740 kg for Zenit-2). It required a new launch vehicle, which was developed at OKB-1 by 1963. That rocket was based on the R-7A ICBM with a new, more powerful, third stage (U.S.

code name: SL-4). It was also used to launch later missions of the Zenit-2 and Zenit-2M.

Apparently the concept of a universal spy satellite with a combination of cameras and electronic sensors was rejected due to the short mission duration. In the mid-1960s, that role was transferred to smaller specialized satellites. Hence, Zenit-4 did not carry signals intelligence equipment.

Zenit-4 was the last photoreconnaissance satellite project developed by the OKB-1. In 1964, the Zenit and R-7 programs were transferred to the Branch #3 of OKB-1 under Dr. Dmitry Kozlov. Branch #3 in the city of Kuibyshev (currently Samara) had its own large factory and was responsible for serial production of OKB-1-designed missiles and spacecraft. That enterprise in 1974 became independent from the OKB-1 and is known now as TsSKB Progress (Central Specialized Design Bureau).[44] Later models of Soviet photoreconnaissance satellites and most of their launch vehicles have been developed by the TsSKB.

The first Zenit-4 launch presumably occurred on November 16, 1963 (Kosmos-22). According to Western observers, the next satellite of that type was not launched until 1965 (Kosmos-59).[45] At the same time there were probably three Zenit-4 missions in 1964 (Kosmos-30, 34, 45). Those satellites are identified in the West as photoreconnaissance vehicles but they are not in the Zenit-2 launch log and are therefore probably Zenit-4s.[46] A modified version, Zenit-4M, began regular service in 1968. Another model—Zenit-4MK—was introduced at about the same time. The latter was probably used for mapping.

No data is available on the missions and success rate of Zenit-4 and its variants. It seems that until 1969 Zenit-4 was launched approximately at the same rate as Zenit-2 (10–13 missions per year), using the same orbital parameters. In 1969 Zenit-4 launches reached a peak (17 missions), while Zenit-2 was in decline (9 missions). The last Zenit-4 was probably launched on August 7, 1970 (Kosmos-355), for a probable total of 74 missions.

It is possible, however, that 1970 was not the termination point for the Zenit program, but rather a time of operational deployment of more advanced satellites based on the Zenit-4 design. Unlike the CORONA program, which ended in 1972 and was replaced by an entirely new system, the Zenit program continued, with increasingly more capable satellites.

CONCLUSION

It is difficult to overestimate the role of space reconnaissance in achieving a peaceful conclusion of the Cold War. Despite numerous technical differences,

Zenit, CORONA, and their descendants met design objectives as reliable reconnaissance systems of an entirely new dimension. As a result, the Open Skies concept has become a reality of international relations in the past thirty years. Space reconnaissance systems became an important political instrument of both superpowers. Constant monitoring from space allowed the Cold War rivals to learn and verify the real military potential of the opposite side. That eventually laid a foundation for negotiations on reduction of strategic and conventional forces. In the Cold War atmosphere of deep mutual fear and distrust, space reconnaissance remained one of the few channels of objective information which was in high demand for keeping the world away from a nuclear catastrophe.

Part Two

VOICES OF THE CORONA PIONEERS

As mentioned in the introduction, the catalyst for this volume was the May 23–24, 1995, conference, "Piercing the Curtain: CORONA and the Revolution in Intelligence." The event was cosponsored by the Center for the Study of Intelligence of the Central Intelligence Agency and the Space Policy Institute of George Washington University's Elliott School of International Affairs. The conference provided the opportunity for the first unclassified discussions of the origins and evolution of the CORONA program and of its impact on the craft of intelligence and on national security policy—the major themes of this book.

While the editors wanted to put together a volume that went well beyond a printed version of the proceedings of the May 1995 conference, we also did not want to lose the historical insights and first-hand perspectives of the conference speakers, most of whom were members of an informal group known as the "CORONA Pioneers." What follows is our attempt to accomplish that objective.

We have extracted the speakers' remarks from the two days of conference discussions which we judged of particularly lasting interest. We then arranged those remarks in narrative form to create a coherent account of the topic under discussion. Each speaker reviewed the remarks we had selected; in some cases they revised or extended them for this book.

What this portion of *Eye in the Sky* represents, therefore, is an overview of CORONA and its impacts from those most directly involved. Their perspectives are crucial to a full understanding of the undertaking.

8
CORONA AND THE U.S. PRESIDENTS

The CORONA program's primary purpose was to gather strategic information for American presidents and their advisors. President Dwight D. Eisenhower's need to understand what was going on behind the Iron Curtain led him to approve the nation's first reconnaissance satellite program.

In this chapter, three former presidential advisors who were instrumental in helping Presidents Eisenhower through Nixon understand and use CORONA intelligence recount the influence that this reconnaissance system had on presidents and their staffs during the Cold War. They stress that each presidential administration learned to rely on CORONA to help assess world events and make strategic decisions during a very tense and difficult time in global relationships. The first advisor, General Andrew J. Goodpaster, served as President Eisenhower's Defense Liaison Officer and Staff Secretary during the conceptual development of the CORONA program and its early operations.[1] Goodpaster is explicit about Eisenhower's regard for, and understanding of, the importance of intelligence and the role it played in determining strategies. He notes:

Eisenhower's experience at the Battle of the Bulge, where the Germans secretly amassed a major force unbeknownst to Allied intelligence, deeply impressed upon him the value as well as the limitations of intelligence, together with the dangers of being caught off guard. Although Eisenhower knew how important intelligence was, he also knew how incomplete it could be. However, his overall understanding of intelligence made him especially sensitive to the U-2's critical role in maintaining America's national security.

Early in Eisenhower's administration, I would say around 1953 or 1954, he considered some proposals to construct high-altitude aircraft and satellites for the purpose of gathering intelligence. People believed that a reconnaissance satellite, or other type of

President Dwight D. Eisenhower examining the American flag flown aboard Discoverer XIII. The flight of Discoverer XIII in August 1960 helped to establish the cover story that the missions were for testing of biomedical and engineering systems. One week later, Discoverer XIV returned the first reconnaissance images of the Soviet Union. (Photo courtesy of the Dwight D. Eisenhower Presidential Library)

high-altitude reconnaissance vehicle, with a high-resolution camera and lens, was quite feasible. The work of Din Land, John von Neumann, James Killian, and George Kistiakowsky, among others, had made such a proposal seem quite feasible. The concept for the U-2—the high-altitude reconnaissance plane—evolved because of this belief.

When the Soviets orbited Sputnik in October 1957, it was quite apparent that something had to be done. So American planners placed a high priority on trying to launch some kind of American satellite so that we could regain confidence in our country's abilities, particularly our national security capabilities. Sputnik was essentially a wake-up call for Americans. The United States suddenly seemed vulnerable to possible Soviet attack. So we shifted our priorities and started stressing the development of satellites.

It is important to note that despite CORONA's initial failures, Eisenhower always said, "Let's not worry about it. Let's stay with it. It's so important, and we need it. We need to just keep going with it." One of the reasons he wanted to continue with the program, despite its initial failures, was because he had seen how beneficial and clear

President Dwight D. Eisenhower with his son John during a television address to the nation following the downing of Gary Powers's U-2 reconnaissance aircraft in May 1960. To President Eisenhower's right is a U-2 image of San Diego. CIA photo-interpreters prepared numerous images of both U.S. and Soviet strategic facilities to demonstrate the capabilities of the U-2. Eisenhower rejected all but one of these images for his presentation, preferring not to demonstrate the intelligence coup achieved by the U-2. (Photo courtesy of Dino Brugioni)

the U-2 photos were. The first U-2 photo he saw was an image of Dallas or Fort Worth as I recall, and you could see the dividing lines in a parking lot on that photo. Because of that photograph, it was immediately apparent to Eisenhower that high-altitude reconnaissance imaging was an instrument of tremendous power.

One of the reasons Eisenhower authorized the CIA to oversee the U-2's operations was because he thought that U-2 flights could be considered a violation of Soviet airspace, a provocation, in other words, and he did not want that provocation made by military people in uniforms—it had to be carried out on the basis of plausible denial. In other words, he was very aware of the risks involved and weighed them very carefully.

Finally, it's important to mention that President Eisenhower had a great respect for photo interpretation and analysis. He liked the degree of confidence it provided and found it very valuable for fighting off a lot of the crash proposals that people wanted to initiate in order to supposedly close the "missile gap." Those types of proposals were

President John F. Kennedy examining the film-return bucket from Discoverer XIV. Kennedy continued the American photoreconnaissance satellite program and increased its scope. He classified all military space launches in order to protect the reconnaissance program from public discussion. (Photo courtesy of A. Roy Burks)

absolutely anathema to him, and when anybody came in with such a proposition, I'd try to warn them that they'd better have their flak suit on because as Eisenhower usually told them, "Don't come in here and try to force me to accept this proposal now. Where the hell were you six months ago when we could have initiated this program on a more orderly basis?"

Like Eisenhower, President John F. Kennedy had a keen interest in CORONA intelligence and relied on it heavily when making strategic decisions. One of the men who helped Kennedy assess CORONA's information was General W. Y. Smith, a member of Kennedy's White House National Security Staff.[2] As Smith explains, CORONA imagery played a vital role in reassessing the strength of Soviet long-range missiles in September 1961 and in estimating the size and capabilities of Soviet forces during the Berlin crisis of October that same year. He also believes that it was President Eisenhower who led us into a whole new era with CORONA. For example, as Smith notes:

Eisenhower deserves a lot of credit for the program because it was his understanding of the value of intelligence, his willingness to take the risks to get that intelligence, his acceptance of CORONA's initial failures in order to gain an eventual payoff, and his ability to use intelligence once he got it, that made the program a reality.[3]

CORONA imagery quickly came into play during the Kennedy administration because of the big dispute that was going on about what size the United States' strategic forces should be. For example, one of my colleagues at the White House, Carl Kaysen, was having this big discussion with Secretary of Defense Robert McNamara on the subject. The argument was about how much we should escalate in 1963 and 1964, a period when both the Polaris and the Minuteman would proliferate. So Kaysen wrote this memorandum to [National Security Adviser McGeorge] Bundy that said, "You know, McNamara's numbers are too high." Kaysen suggested that the figures in the National Intelligence Estimate 11-8-61 indicated a much lower Soviet force level than McNamara thought there was; 11-8-61 suggested a much slower rate of buildup than we had projected. That conclusion not only raised the question of whether our target figures were too high, but also, and more important, suggested that if we escalated our weapons systems, it might trigger a Soviet weapons buildup. So Kaysen used CORONA data to try to hold the Minuteman down to 1,000 and the Polaris down to a smaller number than the Navy wanted.

A couple of months later, in October 1961, when the Soviets cut off our road access to Berlin and began harassing us in the air, CORONA once again came into play. For example, Bill Kaufman, who was with RAND and had been working at the Pentagon on a counterforce study on the Soviet Union's forces, sent an overly confident strategy memo to Bundy, with Kaysen's help, which stressed: "You don't need to worry about Berlin. We've got the Soviets over a barrel—just look at these figures. So take any action you want against them." But Bundy thought twice about the memo. Fortunately, he consulted CORONA intelligence before making a recommendation to the president and realized that Kaufman and Kaysen's suggestion was a bit too ferocious. Consequently, CORONA data strongly influenced the way that White House advisors and many American defense strategists viewed the Berlin crisis. CORONA gave them some confidence in their estimates and decisions.

However, I would also like to note that sometimes we relied on CORONA's data too much. I guess part of my view on this subject stems from my observations about Vietnam, where I think that we mistakenly believed that if we could see enemy targets and count them, we understood their strength and our objective. Nevertheless, we found out that wasn't the case at all. Just seeing and counting didn't tell us the whole story. I think that although photography is very helpful, it is not the only solution. You need to back it up with other things like HUMINT [Human Intelligence].[4]

The third and final advisor, Richard Helms, served as Director of Central Intelligence (DCI) from 1966 through 1972 and was a key figure in determining how CORONA intelligence affected national security plans. Helms was consequently privy to many of the decisions that Presidents Johnson and Nixon made using CORONA information.[5] Interestingly, Helms notes that President Johnson, unlike Eisenhower, did not have much interest in intelligence and only gradually learned to use it. Nixon, on the other hand, realized its importance

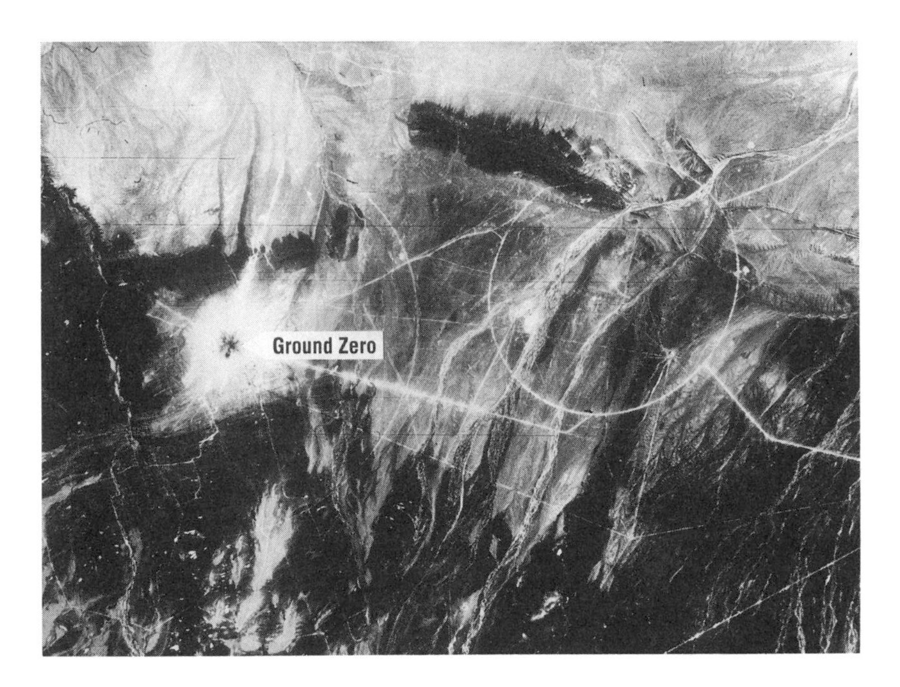

The nuclear test site at Lop Nur, China, on October 20, 1964. CORONA imagery allowed the United States to announce that China was about to detonate an atomic device, undercutting the propaganda value of the event for the Chinese. (Photo courtesy National Photographic Interpretation Center)

from the beginning and used it to negotiate an important part of the SALT I treaty. For example, Helms believes:

When President Johnson first came to office [in November 1963], he was not very interested in intelligence. However, [DCI John] McCone thought that he could convert Johnson to intelligence, so he used to go down to the White House every morning and brief the president. The briefings only lasted about two weeks because Johnson put an end to them.

It would be quite some time—until the Six-Day War in June 1967 to be precise—before President Johnson would finally realize that intelligence had an important role to play in foreign affairs. Why the sudden change? Well, the CIA had correctly predicted how long an Arab-Israeli war would last, and Johnson was impressed with our prediction. The agency hadn't predicted the conflict down to the hour, mind you, but it had correctly predicted that it would occur and how long it would last. So Johnson liked the way we had handled the situation and called me a little bit after the crisis to thank me and ask me if I'd join his weekly lunch group. Now I've mentioned this fact because I want to tell you a story about the luncheon group that is very relevant to the CORONA program.

On August 21, 1968, the Chief of Current Intelligence [CCI] walked into my office and said that the Politburo was meeting in Moscow. Now that really wasn't a secret because it was on the UPI [United Press International] wire. However, since everybody in the Politburo was supposed to be vacationing at the Black Sea that August, we were troubled. We were also worried because we had known for days that the Warsaw Pact forces were maneuvering in East Germany. So the CCI and I put our heads together and figured out that the Warsaw Pact forces were probably going to invade Czechoslovakia.

At 12:30 P.M., I went to the luncheon group at the White House and told them that the Warsaw Pact forces were probably going to invade Czechoslovakia. In fact, I stressed that we were almost sure of it. To support my claim, I started giving some statistics that I had compiled about the Soviets' maneuvers, but President Johnson interrupted and said that the Politburo wasn't meeting to plan an invasion of Czechoslovakia. He claimed that the Politburo was meeting to talk about us. Well, I hadn't the faintest idea what he meant, so as soon as the lunch was over, and we had picked the Vietnamese bombing targets for the day, I cornered Thomas Johnson, President Johnson's note-taker and the current head of CNN, and asked what the heck was going on, and he said, "I'll tell you if you promise not to say anything." I assured him that I wouldn't, and then he told me that we were planning to meet with the Russians the following day to prepare for arms control negotiations. He told me that President Johnson hadn't said anything because he didn't want any leaks.

Well, despite President Johnson's hunches, it seems that my notions about the Politburo were correct, because when my beeper went off, and I rushed down to the White House for a 9:00 P.M. emergency meeting that evening, I learned that the president had had a visit from the Soviet ambassador. He had come to inform us that the Warsaw Pact forces had invaded Czechoslovakia.

Now, back to CORONA. One frustrating aspect of the program that is important to mention was that it took a long time to get the film from the Pacific Ocean to Eastman Kodak for processing, and then to NPIC for analysis. Consequently, by the time we usually got the photos, it was sometimes too late. During the aforementioned Czech invasion, it wasn't until we got the photos that we discovered that the Russians had put large white crosses on all of their invasion equipment in order to distinguish their equipment from the equipment that the Czechs had gotten from them years before. If we would have had real-time readout at that time, we would have known exactly what the Russians were doing the minute they put those white crosses on their equipment.

A somewhat similar problem occurred during the Six-Day War. About the third day of the war, when the Israelis were taking over the Golan Heights, President Johnson turned to me and asked me how accurate the information was that we were getting, and because we were still only getting daily coverage at that time—instead of hourly coverage—I half-jokingly said that it was very accurate as long as the Israelis were still winning. When Dean Acheson heard that, I thought he was going to explode with laughter. Obviously, although CORONA was highly beneficial, it still had certain limitations.

Finally, I'd just like to mention that when it came time to negotiate the SALT I Treaty of 1972, CORONA was indispensable. President Nixon, for example, had told me, "If you can't verify an arms control treaty, we're not going to hold any arms control negotiations." Well, although I didn't believe that any analysts in those days felt so secure in their figures that they were willing to say, "Of course we can verify everything," I still felt confident enough to push forward. So off we went to the arms control negotiations. One of the pivotal points of the negotiations rested on mutual verification; we were trying to match the Soviets' figures with ours. Fortunately, the Soviets said that they'd agree to use our projections about their forces, which we had, unbeknownst to them, compiled using CORONA. So, in other words, CORONA produced the arms control statistics that we used to resolve the SALT I negotiations.

Overall, CORONA provided Presidents Eisenhower through Nixon with vital strategic intelligence. Each administration learned to rely on and value the information the program furnished. CORONA supplied the Cold War presidents with the type of information that they needed in order to assess and defuse a variety of dangerous situations during a very volatile period in world history.

9
THE ORIGIN AND EVOLUTION OF THE CORONA SYSTEM

In this chapter, five CORONA pioneers recount the various roles they played in the development of America's first reconnaissance satellite system. Each one shares his own unique perspective about the origin and evolution of the CORONA system by detailing the work he did as a government, military, or industrial representative. Their comments range from the original conceptualization of CORONA to the various technical aspects of designing its cameras and lenses.

Although each pioneer concentrates on a different aspect of CORONA's development, one common theme that emerges from all of their discussions is the idea of "teamwork." Collectively, they agree that CORONA would not have been possible without the extraordinary cooperation that occurred between the government, the military, and industry. As the pioneers explain, in the late 1950s and early 1960s, when tensions between the Soviet Union and United States were extremely high, all concerned sectors recognized the pressing need for satellite intelligence and worked together to make sure CORONA came to fruition.

One of the pioneers, John N. McMahon, who worked with the CIA/Air Force team that developed and oversaw the U-2, A-12, and CORONA programs, begins the discussion with an appreciation for just how daring the people were who worked on CORONA during the project's nascent years.[1] McMahon stresses that the program's pioneers must have possessed an incredible amount of courage and determination in order to continue pressing for CORONA's development despite its numerous early failures. He also recounts why politicians and the Intelligence Community believed that CORONA was so vital to American security. McMahon asks us to try to think of what space was all about 37 years ago [1958]. At that point in time we had no experience in space and to think of orbiting the earth was beyond the comprehension of the average person. Think of the courage that a handful of government officials—military and civilian—had to possess to seek approval to carry out our first venture into space. Think about the bold and courageous efforts by contractors to do what no man or woman had done before, and the anxiety of the government officials who had the vision to persevere in spite of the many setbacks and failures. Only the stoutest hearts could have prevailed. Could we orbit a satellite? Could it take meaningful pictures from space? And could we really bring it back in a film-recovery capsule? All of these were very profound questions that required actions and results—the first of their kind in the annals of the United States.

Why did so much effort and risk go into the first satellite effort? Just stop and think of the times. Remember that our lack of knowledge of the Soviet Union prompted President Eisenhower to proceed with the extraordinary program of the U-2 overflights, even though he recognized that it would only be a year or two before the Soviets would counter those flights. So he pressed for alternatives, and they took the form of the A-12, now known as the SR-71 reconnaissance aircraft, and a satellite system under a projected Air Force program known as the WS-117L.

The crying need to understand what was happening in the Soviet Union drove this program and gave tolerance to the risk. There was a general concern in this country that there was a missile gap and that we were falling behind the Soviets in our military posture. Our national security was indeed at stake. Also, bear in mind that President Eisenhower advocated "Open Skies" in 1955, and when Khrushchev turned him down, he could only conclude that the Soviets wanted to conduct their actions in a closed society and in secret from the world. What also added stimulus to the CORONA effort was a proclamation by the Soviets that they had fired an ICBM in August 1957—and, of course, the Sputnik satellite also orbited in October that same year.

And so the CORONA program moved forward. It proceeded with great haste. In April 1958, with presidential approval, CORONA came to fruition. In the span of a little over a decade, it produced 145 launches at a cost of $850 million and returned 300 million nautical miles of cloud-free photography revealing all of the Soviet missile complexes—medium, intermediate, and long-range; each class of their submarines; a complete inventory of their bombers and fighters; the Soviets' ABM effort; and their weapons storage. It also gave us maps for strategic targeting. More important, it gave our policymakers the confidence to enter disarmament talks and fashion the SALT I Treaty. It was a remarkable program.

General Lew Allen Jr., another CORONA pioneer, was one of the key Air Force personnel involved in the early satellite reconnaissance program. Allen saw the CORONA project develop out of the Air Force's WS-117L program and the RAND Corporation's early space studies. He also served as Director of Air Force Special Projects during the last several years that CORONA was in operation.[2] In his discussion, he recounts how the project evolved from WS-117L

Technicians at work on the Agena vehicle for Discoverer XIII. After a string of disappointing failures, the program managers decided to launch Discoverers XII and XIII as instrumented test vehicles, without cameras. Discoverer XIII's payload was an American flag stored inside the reentry vehicle, the first object to be successfully retrieved from space. (Photo courtesy of the U.S. Air Force)

and the RAND studies, and how a handful of key Air Force officers worked to convince the Air Force high command of the benefits of a satellite system. In order to understand the origin of the CORONA system fully, Allen believes that

> we need to go back a little bit and remember that [Theodore] von Kármán in 1945 wrote a report called "Toward New Horizons," which was very influential in setting the tone of the Air Force research and development program for the next fifty years. In that document, he observed that if you had the propulsion for a long-range ballistic missile, you were very close to having the propulsion to place something in orbit. But there were no speculations in that report as to why one would want to do that.
>
> RAND got started as a result of a very prescient early document which observed that there were uses for satellites and that reconnaissance was certainly a key one of those uses. As RAND formed, and those studies continued, that whole concept was expanded and fleshed out to where, by 1953, there was a substantial body of studies by RAND that observed that there were a variety of things that could be done in space. And this volume of studies, because of the sensitivity of reconnaissance missions, was

classified Top Secret. It was hard to take the studies seriously when there was no rocket boosting capability. But as boost capability began to be apparent in the early 1950s, and we discovered one could and should build an ICBM, then I think interest in RAND's ideas picked up.

Around 1954 or 1955, there was a project office established at Wright Field under Bill King that went around the country and presented the RAND report and suggested that the Air Force should begin moving toward implementing some of the things that RAND had observed were possible. But Bill did not always receive a warm reception. Many people were very skeptical of these space concepts. Bill says that he used to carry a dishpan with him in which he'd place a ball bearing and swing it around to try to show how forces could keep an object in orbit. But SAC [the Strategic Air Command] was still very uninterested and [General Curtis] LeMay basically kicked him out. You know, LeMay was a very pragmatic person. But because nothing had flown in space, he was skeptical. He wasn't very interested in it at the time. This was pre-Sputnik.

By 1956, these studies led to the establishment of WS-117L, but without any significant amount of money attached to it. The system was based on the West Coast largely because as the project was established, [General Bernard] "Benny" Schriever and Simon Ramo observed that it would probably interfere with the ICBM/IRBM programs that they were conducting on the West Coast, unless they also had control of WS-117L. So they wanted it moved out there and integrated in some fashion with the missile development programs. Obviously, that turned out to be a good decision.

In 1957, after Sputnik flew and people began to take space seriously, the dollars began to flow and WS-117L actually began to get funded. The Air Force's interest grew greatly. I can recall that shortly before Sputnik, we were all encouraged *not* to use the word "space" in any documents, but immediately after Sputnik we were encouraged to use the word "space" in *all* documents. SAC began to believe that it was really possible to have objects in orbit, and it began to recognize that space reconnaissance was extraordinarily important for target planning. So they really began to give it support. But, by the end of 1957, it was clear to most people, I presume even to the 117L people on the West Coast, that the schedules were unrealistic. 117L was trying to address a broad menu of requirements including both the Atlas and Thor programs. It lacked a focus.

In the latter part of 1957 and the early part of 1958, the decisionmaking process moved to Washington. The president, his advisors, and the Director of Central Intelligence all said that we had to significantly change our approach in order to adequately meet the demands of the nation. So, when those decisions were made, we decided to pursue CORONA as a covert program. The Air Force people recognized that overflights were politically sensitive, and they consequently acknowledged that a covert program was the only way to really be successful.

As James W. Plummer, the former manager of the Lockheed Aircraft Corporation's CORONA development team, remembers, CORONA's covert nature heavily affected everyone involved with the project.[3] Plummer recalls that it required a tremendous amount of diligence to keep CORONA secret. He

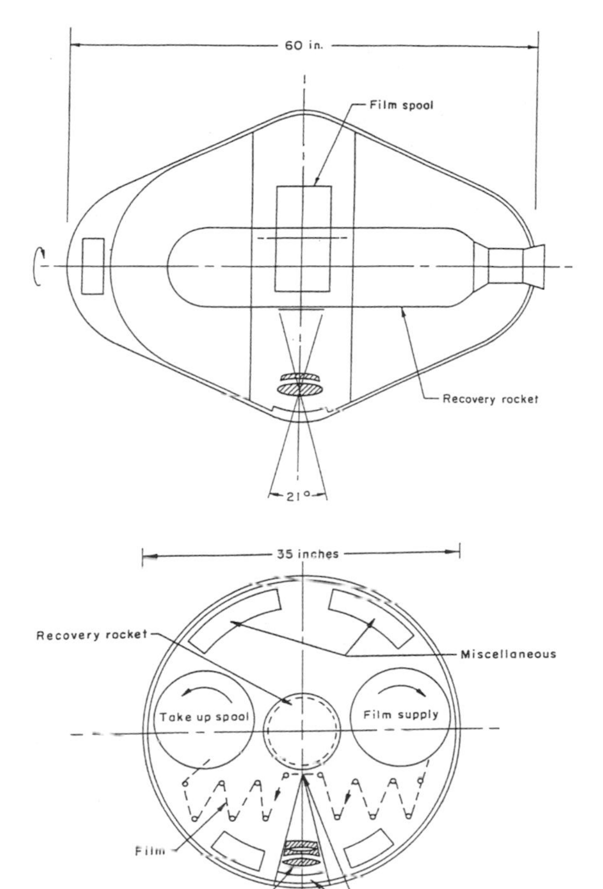

The side and front views of an early spin-stabilized film-return reconnaissance satellite proposed by RAND. The design was simple, but did not offer much room for growth. It was pursued for a short time at the beginning of the CORONA program before it was abandoned in favor of a new camera design proposed by Itek.

himself sometimes even had to lie to friends and superiors in order to conceal the system. For example, Plummer notes:

The existence of the CORONA program was highly classified and had to be denied. We could not admit the existence of the program to anyone, including superiors in our own places of work. It was a truly covert program. There were many, many steps taken by the CIA and the Air Force which allowed us to stay totally submerged—such as having a separate accounting system. Our monies were all hidden in the system. And of course, the word "CORONA," or even the letter "C," which scared us all to death, was never used either. We were very, very conscious of that in everything we did. You would not talk to someone unless you really knew that person was part of it. It was a totally covert system based on a strictly "need-to-know" basis.

Because of the secrecy, Colonel Lee Battle, who was the System Project Officer, and I had difficulties facing some people in the outside world. For example, when we were launching the mice as a cover story during the early days, we would hold press confer-

Colonel Lee Battle was the Air Force officer initially in charge of CORONA operations in Los Angeles. Battle was instrumental in getting CORONA operational. (Photo courtesy of James Plummer)

ences—with the Air Force's permission, of course—and these smart reporters would come into the room and start asking questions which we couldn't answer. It was hard to stonewall them because they wanted to bore in, especially since they suspected we were doing more than we said we were.

Another situation that was difficult was when smart engineers would come and knock on your door and say, "I want to talk to you about an idea I have." And what they would do is reveal their own private design for something that might be able to be done in orbit. But then you had to turn them off by saying, "Well, it's a pretty good idea, but we haven't got time for that," or you'd have to find some other way to get them out of your office.

One final security matter that was difficult was when we went to the Pentagon to report to [Undersecretary of the Air Force and Director of the National Reconnaissance Office] Dr. [Joseph] Charyk and to the people at the CIA. We generally would go around and see people in the Pentagon who knew we were doing something and they were generally pretty high-ranking officers. This was very difficult because it was hard to be evasive to a high-ranking officer. General LeMay was a good example. He was very, very difficult to tell a story to and not tell the whole story. There were also others who knew that something was going on and that thought the explanations they had gotten out of this colonel or that Lockheed engineer were very bad answers to their questions. This led to great problems.

In addition to CORONA's secrecy, another element of the program that impresses General Allen was the enormous amount of cooperation that developed between the CIA and the Air Force during the early years of CORONA. Allen remembers:

There was an enthusiasm on the part of the people involved to really make this thing work. The Air Force/CIA team pulled together extremely well. Although ARPA [Advanced Research Projects Agency] was nominally in charge, it was largely just a pass-through organization. The activities were done in key measure by the program officer—Richard Bissell at CIA, and his deputy on the West Coast, [General Osmond] "Ozzie" Ritland. The project seemed to work well because there was a very clear mission and a clear understanding of that mission. There was no ambiguity throughout the life of CORONA. It was aimed at national intelligence, intelligence that national strategic planners and the Air Force needed.

The initial cooperation between all parties was remarkable, even though the delineation of responsibilities between the Air Force and the CIA was not always clear. It was very fuzzy, apparently even to Bissell and Ritland, and was worked out primarily by the people on the West Coast: Lee Battle, who was the program director, and Charlie Murphy, who worked at the helicopter facility at Palo Alto. They did an extraordinary job pulling it all together without squabbling. There were some squabbles, of course; it is hard for porcupines to mate without there being a few sticky points, but those squabbles were handled over the life of the program in a quite satisfactory manner.

Richard L. Garwin, who was a member of the President's Science Advisory Committee and the Defense Science Board during the CORONA era, recounts CORONA's origin and evolution from his unique perspective as a member of the Land Panel—a group chaired by the renowned inventor and scientist Edward "Din" Land, which was responsible for recommending what kind of technology to develop in order to guard against a Russian nuclear attack on the United States.[4] Although Garwin did not join the Land Panel until after it had recommended the development of CORONA to President Eisenhower, his story still provides some unique insights into the panel's operations and culture and its influence on the continuous evolution of the CORONA program. Garwin recalls:

Edwin H. Land, inventor of polarizing film and the Polaroid instant film process and camera, and president of the Polaroid Company, was both a genius and a showman, and an extremely productive and forceful person. Land became chairman of the Technological Capabilities Panel's (TCP) Project III, an activity organized in July of 1954 by James Killian in response to Eisenhower's challenge to use new technologies to counter the Soviet potential of nuclear attack on the United States by intercontinental bombers.[5] Projects I and II dealt with continental defense and striking power.

By November 1954, President Eisenhower had approved the TCP's Project III proposal to build the U-2. The responsibility for its development and operation were assigned to CIA. It was to be covert and subject to special security procedures. Its high-resolution, small-format camera constituted a new optimization of aerial photography by allowing the capture of an enormous amount of imagery.

In early 1955, Richard Bissell, CIA director of the U-2 CIA/Air Force project, asked Din Land and the TCP to advise him on systems for the technical collection of intelligence, which included the U-2 and the A-12 (which would later be known as the two-seat SR-71 and produce the Mach 3 reconnaissance aircraft). Bissell later received word from Land that he was to develop the CORONA system. Din Land's intelligence panel continued its activities and constituted an important part of the nation's strategic reconnaissance program. It was involved in the continuous improvement of CORONA as well as successor satellite systems.

I was pleased to be able to join the Land Panel in 1962, or maybe somewhat earlier; I don't recall the exact date. Most of the recommendations of the Land Panel involved the National Reconnaissance Office—the NRO—for which the CIA and the Air Force were the two chief implementing elements. However, there were also non-NRO recommendations dealing with R&D [research and development] that were not specific to a particular satellite system but were of general use in improving the capability of any system. For instance, improved image storage, photographic interpretation techniques, dissemination of imagery, and other intelligence data were all nonsystem-related.

During my years with the Land Panel, we met typically several times a year for two or three successive days, either in Washington or in the field. In either location we would hear from government sponsors about the recent performance of systems in operation, look at the development of new systems, and hear proposals for more advanced systems. We shared in much of the anguish that accompanied these programs during their early days and also in some of their joys of success. Occasionally we would meet at a contractor's facility and sometimes in the magnificent boardroom at Polaroid where Din Land's cook might provide us with dinner, although as I recall sometimes at 11 P.M. Once our Land Panel meeting did not adjourn until dawn. That was unusual, but it was not unusual to work until midnight and then to be invited to Din Land's personal laboratory for a half-hour demonstration of his latest work on color vision.

The Land Panel would also sometimes meet in the Old Executive Office Building next to the White House, where we reviewed programs and proposals, and looked at some of the most recent images. Din Land inspired us and kept us on track. Our job was not primarily to invent solutions, because there were usually plenty of those to exhaust budget resources. Rather, our task was, as quickly and surely as possible, to separate the wheat from the chaff and to encourage the wheat. We tried hard to increase the research and development effort in the Intelligence Community in nonsystem areas, and occasionally, in addition to helping to choose between various systems, we did contribute valuable technical innovations.

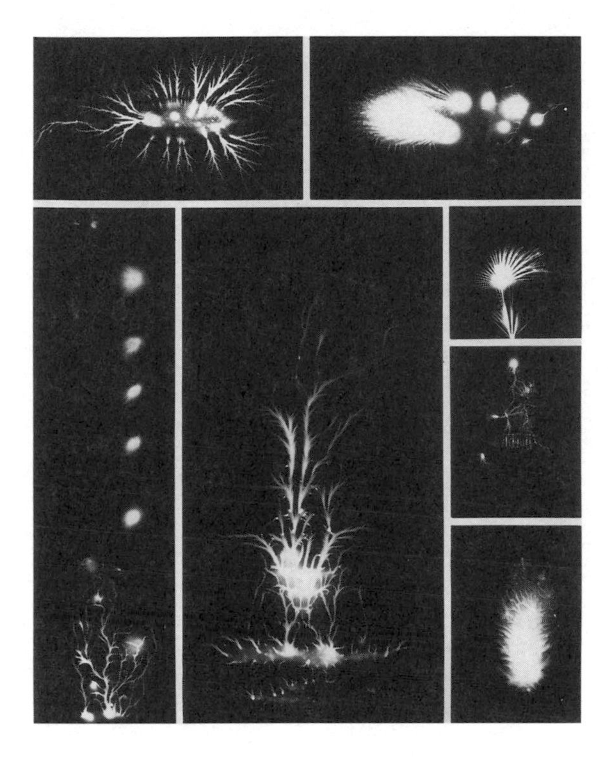

Examples of the "corona" film problem encountered in orbit. The camera's film rollers built up static electricity and discharged, exposing the film. The usual result was film fogging, but occasionally the phenomena produced the spectacular patterns seen here. The problem was solved through a methodical trial-and-error process. (Photo courtesy of A. Roy Burks)

The Land Panel often worked at a very technical level, and some of these technical considerations had a major impact. For instance, during the mid-1960s, when Donald F. Hornig was Presidential Science Advisor, the Land Panel undertook a study of the potential performance and utility of the Manned Orbiting Laboratory program for strategic reconnaissance—which was the primary, although classified, goal of the program. We arrived at the very definite conclusion that humans in space would have harmful effects on strategic reconnaissance, and this report was presented by Don Hornig to Secretary of Defense Robert F. McNamara. It was instrumental in the demise of the Manned Orbiting Laboratory program.

In a draft of a Land Panel report on March 1, 1965, I tried to reflect the panel's philosophy. "The panel is very much disturbed that continuing low-level research is not performed at the necessary scale in this field. Conservatism is certainly a virtue in the later stages of development of an operational system, but if it is to be practiced there, then there must be adventuresome exploratory development, an advanced development program, and a short feedback cycle to address problems that appear during the development phase."

These strategic reconnaissance programs were often characterized as wasteful and inefficient, if not worse. But that was not so for most of them. I was extremely pleased with the quality of the leaders of the technical programs and their willingness to listen to advice and criticism from the outside—and then do something about it.

James Plummer (*left*) was Lockheed's program manager for CORONA during its early years. Here he is after being thrown in the swimming pool during the celebration for the success of Discoverer XIII. Also shown are Aub Grey (*center*) and Bill Wright. (Photo courtesy of James Plummer)

Like Garwin, Plummer traces CORONA's origin and evolution from his own perspective. As previously noted, Plummer was the manager of Lockheed's CORONA development team. He discusses here how Lockheed recycled its early work on the WS-117L for use in the CORONA program, traces his role in designing CORONA's payload system, and recounts the steps he had to take to insure the program's secrecy. He also recalls the difficult schedule required to keep CORONA on time, and how he and many of the CORONA pioneers viewed the program's early failures. Overall, Plummer feels

fortunate to have been at Lockheed during the early days leading up to the WS-117L program, and to have had the opportunity to join the small group of people working on the final draft of the Lockheed proposal. Our proposal was not very thick or detailed by modern standards, but at least we won the proposal. That contract marked the beginning of the space group at Lockheed.

In 1956, it was kind of a stop-and-go situation because there was a great deal of concern throughout the entire system as to what was really to be done. Could it be done? How much was it going to cost? What kind of schedule were we going to have, and where would the funding come from? So it was the typical thing of plan and replan. There was not much physical progress during that period.

Still, at approximately the same time, a small group of engineers working in this small division at Lockheed, which was originally located in one of the research labs in Palo Alto, designed a launch vehicle. This vehicle would later be called the Agena. So we actually had a mockup of an upper-stage vehicle that you could go look at. We also had pretty much figured out all of the basic technical factors of a readout-based satellite reconnaissance system. We had never done any of this, but we had figured out what had to be done. And that's where I think two individuals deserve a tremendous amount of credit. The two individuals at Lockheed that I give extreme credit to are Willis Hawkins, who was at the time the vice-president in charge of space, and Fred O'Green, who was essentially the program manager of the WS-117L.

When Sputnik was launched, all of a sudden satellites became a big issue for the Air Force and the country. We essentially received total funding to go and create the system as fast as we could. We started making a great deal of progress.

Then, in 1958, at a time when I was responsible for the payload end of the WS-117L program, I got a big surprise. I was summoned to my boss's office and he said, "Jim, are you willing to take on a new job?" And although I was kind of disappointed because I liked the job that I was on, I said, "Sure." And he said, "Well, we want you to go underground." And I didn't know what underground really meant, but he said, "We want you to head up a totally covert program. We want you to disappear from Lockheed and all your friends, and not tell anyone where you're going, and go off—take these three or four drawings" that they handed me "and go off and build that thing." It was a design that had come out of some of the original work at RAND by Amrom Katz and Merton Davies. It essentially called for a large football-like unit with a camera in the middle, which was based on a Fairchild camera that already existed. It also had a retro-rocket in the back and a sphere taken from the Atlas ICBM instrumentation capsule built by GE.

Well, I had to just disappear. I had no place or people to work with. So I went out and rented a motel room up on El Camino Avenue and started to look over these drawings. I was then given authority by my superiors to go out and find a factory. I found a very handy spot in an essentially unused prototype facility at Hiller Helicopters. Hiller was willing to rent that facility to us. It had a shop, offices, telephones, and other things that we needed. So we took it over and told Hiller, "From now on you can't enter the building." It was a totally closed building, and very few people at Lockheed knew that we were doing our work there. My superiors even told me to take a circuitous route from home each day so that people couldn't track me.

Well, we worked on that design for some time and thought that maybe it could be done. Then, all of a sudden, we were confronted by the top leaders, General Ritland, Mr. Bissell—"Mr. B" as we called him—and told that we were going to drop work on the football design and go a new route. It was kind of refreshing, but certainly a shock to us. At that time, they handed us this work statement that was a page and a half of

The Lockheed CORONA engineers celebrating after the success of Discoverer XIII. James Plummer is at right. (Photo courtesy of James Plummer)

written material, together with a schedule, and that was the basis of the CORONA program. All it said was it's going to be a satellite, it should be based on WS-117L as much as you can, and it should be compatible with an overt biomedical payload for cover. Photographs should be obtained at 25 feet or better, and get the maximum ground coverage that you can. We were also supposed to make sure that the film capsules were recoverable within a range of plus or minus 200 nautical miles of latitude and plus or minus 75 miles of longitude. And that was all there was to it. They also specified that we would use Itek and General Electric as the two basic subcontractors.

Then they attached a schedule. The design release was to be accomplished in two months, and the prototype was to be accomplished two months later. The first flight units were to be delivered ten months from the starting date, and the first flight was scheduled for eleven months from the starting date. Well, if we had not had the wonderful team that we did at Lockheed, and the strong teams that we had at Itek and General Electric, we never could have made it.

Only a select few were briefed on CORONA, and that made it a little bit difficult for us because we had to get things done without telling anyone what we were doing, and sometimes without authority. Sometimes we'd even have to go to a really smart physicist or dynamist and say, "Do this for me, but we won't tell you why we want it." It was difficult from that point of view.

It was also difficult because there were a lot of failures. So we had to take them to higher levels in Washington, like Dr. [Joseph] Charyk—who was the director of the National Reconnaissance Office and our primary reporting point—and explain it to

Film-return buckets were wrapped in a light-tight plastic bag before being placed in a steel drum and transported to the United States for film processing. Here a bucket is being removed at the Advanced Projects facility before final shipment to the processing facility. At left is Bill Snyder, Operations Manager. Center is Ken Perryman, Lockheed Reentry Vehicle Test and Integration Manager. At right is A. Roy Burks, who was CORONA's technical manager in the mid-1960s. (Photo courtesy of A. Roy Burks)

him. Fortunately, he was very, very patient and listened to us. He was also a very strong technical person and could keep up with us. But sounding exactly like the prime contractor that I was, we looked at these failures as a bunch of engineering successes. I really do mean that, because there was an engineering achievement to every one of them. There was an achievement even in the failures because we learned to diagnose problems that never could have been done before, especially since this was the first time that a long-lasting vehicle of any kind had to work perfectly without any person there to check it out and bring it back home, as is typical in airplane design. So, it was very, very difficult from that point of view.

Walter J. Levison worked as the Itek Corporation's first program manager on CORONA. His responsibilities included overseeing the development of the CORONA camera system and, in 1958, all of Itek's aerial surveillance. Before joining Itek, Levison served as the assistant director of Boston University's

Physical Research Laboratories—the laboratories in charge of developing the Air Force's specialized reconnaissance programs.[6]

Levison begins his discussion with a brief history of the camera used during the WS-46IL reconnaissance balloon program (CORONA's forerunner), and then delves into the history of Itek and its role in the construction of CORONA's lenses and cameras. He also details many of the technical specifications of the program's other photographic equipment, as well as how he and his fellow engineers solved the challenges and problems presented by the cameras, lenses, and film system. Levison stresses that panoramic cameras

have been around for a long, long time. But they were never thought of as aerial cameras until Dick Philbrick—who was the Air Force liaison officer at the Boston University Physical Research Laboratory during this period—modified a strip camera and put it into an airplane. The camera rotated around the longitudinal axis of the aircraft while the film moved synchronously with the rotation of the camera. It took a very dramatic photograph of Manhattan from an altitude of 10,000 feet and caused quite a stir because you could see the entire island beautifully. The photograph appeared in *Life* magazine.

You could ask the question of why we used panoramic photography? Well, it's very difficult to get wide-angle coverage from a lens and still get high resolution. But with a panoramic camera you just have to cover a very narrow angle and sweep the rest of the picture mechanically.[7] This was the thought that we had when we got the 461L contract, which was the balloon camera that was part of the GENETRIX program. While working on the GENETRIX at Boston University's Physical Research Laboratory, we designed a 12-inch f/5 triplet lens camera. We designed it to be a very simple and lightweight unit because of the constraints of the balloon. We didn't even paint it because we were trying to save every ounce we could. It was essentially a nineteenth-century "circuit" camera design, the type photographers used to photograph long rows of people. We now call it a direct scanning panoramic camera. It was originally designed to be flown at 100,000 feet. There were 40 of them made, but only three were deployed.

As a result of our efforts, for the first time in history an aerial camera consistently produced 100 lines per millimeter. That was incredible quality when you consider that up until that time World War II photography had maybe 10 lines per millimeter, or maybe 15 lines at the most. Just to give you a feel for the quality of these photographs, if you started out with a negative that was two inches wide by 25 inches long, and you enlarged that photograph 20 times, it turned out to be a print three feet wide by 42 feet long, and if you looked at it from one end to the other, it would have appeared sharp to the unaided eye.

Amrom Katz was so impressed by the quality that we had achieved that he actually wrote me a letter, and I still am proud that I have that letter. It says, "I think this is one of the top achievements in the history of aerial photography . . . a truly magnificent achievement."

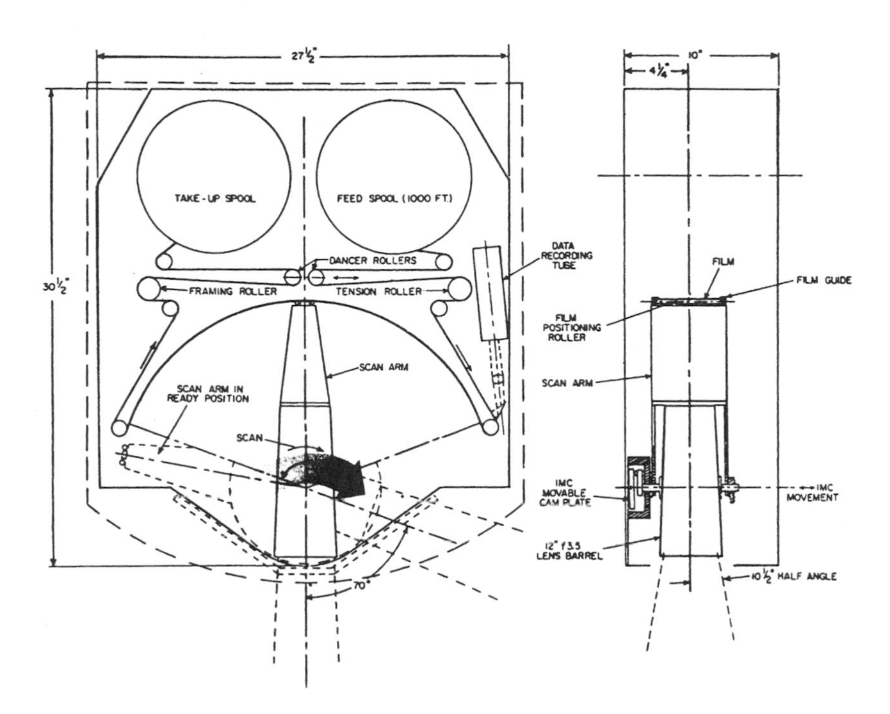

A schematic of the HYAC balloon camera. This panoramic camera served as the basis for the CORONA camera system. For CORONA, the focal length was doubled, the scan angle decreased, and the film-supply cassette and take-up reel moved from above to beside the camera.

Itek Corporation was formed in the fall of 1957 by Dick Leghorn, and on January 1, 1958, it acquired the laboratories at Boston University. Leghorn had long been an advocate of prehostilities reconnaissance and was a major contributor to the Open Skies proposal that President Eisenhower had introduced at the Geneva Summit of 1955. The notion that aerial reconnaissance could be used, as it eventually was, to promote peace and verify disarmament, was one that interested me deeply.

At Itek, we had a contract to build part of the data-processing equipment for the WS-117L. We were aware that the camera was a spin-stabilized 6-inch focal-length panoramic camera—very much like the camera that Philbrick had originally modified and flown over Manhattan. We also knew, of course, about the balloon camera, having designed it. We knew what its potential was. And so we proposed, on an unsolicited basis, a 24-inch Tessar f/5 camera and narrowed the scan angle down to 70 degrees. We thought we could probably get 20–25-foot ground resolution from satellite altitudes and fortunately that proved to be correct. The project office funded both projects for a very short time, but inside of a month—probably due to the influence of Din Land—the project office switched over to the longer focal length camera. Itek became the prime contractor for the camera system; we built the lenses, while Fairchild got a subcontract from us to manufacture the camera.

We were fortunate to have a very small but excellent photogrammetric group at Itek, first under Claus Aschenbrenner and then under Ron Ondrejka. Throughout this entire program, they were very helpful in getting things like horizon and stellar cameras incorporated into the system, in addition to developing equipment to help in photo interpretation. There was a piece of equipment called the Gamma Rectifer, for example, which produced a very large, near-vertical, map-like reproduction of the terrain from these panoramic photographs. It was extremely flexible because it compensated for the earth's curvature, the stereo-angle, image motion, and the pitch and roll of the spacecraft. They also built a very large rear-projection photo-interpretation viewer.

All of CORONA's subsystems had their problems, and the camera was no exception. The most notable camera problem was the film breakage caused by the hostile environment. The acetate film became very, very brittle and had to be treated with the utmost care if it was ever going to survive. Well, it didn't survive. The film constantly broke in orbit and, as a result, the launch schedule came to a grinding halt. The film and the camera would never really successfully work until Eastman Kodak came up with the polyester-based film. That's what really put that particular problem to rest.

Although we were somewhat disappointed with the results of the first successful CORONA mission—it only had about a 35-foot ground resolution compared to the design objective of 20—we were still very encouraged by it. The system had worked, and what's more, we knew that we could improve it. So, we went through a series of improvements. In the long run what we wound up with were two cameras. We changed the lens to a 24-inch Petzval f/3.5 design. Now a Petzval lens, interestingly enough, is a very old lens design. It was used in portraiture. It's what we call a soft focus lens and it suffers from extensive field curvature. But what we recognized, or what Bob Hopkins of the University of Rochester who designed the lens recognized, was that extensive field curvature is not that much of a problem in a panoramic camera because you can put a very lightweight cyclindrical field flattener in the focal plane and get essentially diffraction-limited resolution. So it was this ability to take an old lens design, the Petzval, modify it with a field flattener, and use it in the slit of a panoramic camera, that made even higher resolution possible.

Another challenge that we faced was that the 24-inch Petzval f/3.5 vibrated a lot. To compensate for the problem, we decided to constantly rotate the lens. Fortunately, this adjustment decreased the vibration and we were able to get 180 to 200 lines of resolution per millimeter, which equaled 10 to 12 feet of ground resolution—a major improvement in performance.

The next major improvement was the MURAL camera system, which consisted of two C‴ ["C triple prime"] cameras, one pointing forward, and one pointing backward, at 15 degrees each. This gave us a stereo angle of 30 degrees, and from that you could create a stereo model so that interpreters could calculate—in addition to being able to make the standard measurements in *x* and *y*—measurements in *z*, or in other words, they could get "height" information as well.

The film take-up reels inside a CORONA bucket. Note the two film strips at top.

The final modification of the CORONA camera was the Constant Rotator or J-3. We were still having an intermittent slip of the scan arm with the C‴ lens cell (the system used in the MURAL and J-1 cameras). So what we did was put a different drum on it so that the scan arm, as well as the lens, constantly rotated. This allowed the stereo coverage to work properly. It was the last of the CORONA series and was an extraordinarily successful and reliable system. Statistics show that it was a highly reliable camera system and a highly reliable vehicle.

The ultimate improvement in the CORONA system was when we went to the J-1 and KH-4A's dual film-bucket system. After that innovation we had a great deal of flexibility. For example, we could expose a bunch of film, load it into one of the buckets, and then reenter that bucket while still maintaining the satellite in orbit for another series of film. In other words, the system gave us twice as much coverage as the one-bucket system and the flexibility to photograph a different area than we had for the first bucket. In all, the J-1 increased CORONA's orbital life to approximately 20 days and operated without failure for more than five years.

The final problem I would like to talk about is the so-called corona effect on the film. From the very beginning, we had evidence of static electrical discharge. Sometimes it was dendritic tree-like patterns, other times it was bands of increased density, but we always had corona. Ed Purcell of Harvard, who was a member of the Itek Science Advisory Board and the Land Committee, made several suggestions for improving the situation. We tried them all. We even tried making the film roller out of static-free conducting rubber, but that didn't consistently work. So what we finally did was put the rollers in a test rig; we then put them in a steam bath and installed them in a camera so that we could test them in an environmental chamber. If we got static discharge, we discarded them. But if we didn't, we used them. And that's the way that problem eventually got resolved.

There were many people at Itek who contributed to the success of this camera, but it would be particularly unfortunate if I didn't at least mention two names: one of them is John Wolfe, who was the project manager for most of the CORONA program, and the second one is Frank Madden, who was the chief engineer on the camera from the very early days of the balloons right through to the final J-3 design. Madden's calm guidance, his penetrating intellect, his pragmatic approach, and his ability to come up with solutions to problems that appeared to be intractable were uncanny. He says that he thought the "corona" static problem was resolved by divine intervention. I note that it was his determination that eventually solved that problem.

In summary, as Allen, Garwin, Levison, McMahon, and Plummer have noted, determination was an element that everyone who worked on CORONA possessed. Despite a number of early setbacks, the CORONA pioneers developed a distinct "can-do" attitude that helped them push themselves and make sure their plans became a reality. One of the most remarkable aspects of the CORONA program was the extensive cooperation that occurred between the government, the military, and industry. The teamwork between these three sectors assured the development of America's first reconnaissance satellite. Ultimately, it seems that everyone involved with the program sensed that CORONA would be a key factor in deciding the outcome of the Cold War. But perhaps the best way to summarize the evolution of the CORONA program is to let some of the CORONA pioneers recount their ideas about the system's development, for, as General Lew Allen notes:

We were in a very unstable situation because we did not know what the Russians' capabilities were. Consequently, I think all of the people involved felt very strongly that intelligence was critical and that knowledge of the Russians' strategic capabilities was important in order to achieve stability.

This program contained very challenging things. These challenges were all brand new for everyone involved, and it took an enormous amount of dedication and innovation to pull them all off. As one looks back on the program, it is remarkable that these people persisted even though a long series of failures occurred. I think it's a remarkable story. It was a program of remarkable accomplishment, and remarkable cooperation between the Air Force and the CIA.

James Plummer has similar admiration for the teamwork and cooperation that existed during the CORONA period and notes:

The real work was done by the shop people, the technicians, and the design engineers. So if a person were to say, "Who's responsible for CORONA?" I'd have to say it's measured in the hundreds and hundreds of people. It is obvious that no single individual created this system. Industry worked with government. It was clearly a team effort.

But perhaps Walter Levison most accurately captures the spirit and feelings of the men and women who worked on CORONA:

CORONA was a great program. We worked very hard. We enjoyed being on it, and, it would not have been possible without the kind of environment that Dick Bissell, Joe Charyk, Jim Plummer, and the rest of the management staff provided for us. We all felt we were doing something important. We all felt they trusted us. And we all felt we were being amply rewarded for this job.

10
CORONA AND THE REVOLUTION IN MAPMAKING

CORONA had a major impact on both American military and civil mapmaking during the height of the Cold War. In this chapter, four photogrammetrists and geodesists who worked with the CORONA program discuss how they used reconnaissance satellite imagery to create a global mapping system that defense officials could use to locate and target objects more accurately. They also describe how they used CORONA's photographs to develop methods and technologies for revising civilian maps of the United States and other areas of the world.

These four mapping experts' comments range from their first experiences with the CORONA system to the profound effect that the satellite had on both the military and civilian mapping communities. Although each of them has a different perspective on CORONA, two common themes emerge in their discussions: the innovations they developed in order to exploit CORONA imagery during the mapmaking process, and the tremendous cooperation that occurred between the military, civilian, and industrial sectors when it came to exploiting CORONA's mapping and targeting potentials. In all, these four mapping experts reveal that CORONA fostered a revolution in the sciences of geodesy, photogrammetry, and cartography, a revolution that would lead to the development of a sophisticated global mapping system for both civilian and military use.

Elaine A. Gifford was a photogrammetrist at the National Photographic Interpretation Center (NPIC) during the latter part of the CORONA era. Gifford began her career at NPIC right after college.[1] In her discussion, she gives a thumbnail sketch of the mapping process and recounts the kinds of questions that mapping experts who were working with the CORONA program had to address. Gifford notes:

My involvement with the CORONA system began in 1965 when, as a brand-new employee of the CIA, I started my duties at the National Photographic Interpretation Center. I still remember the awe of it. Things were coming though NPIC's door on a regular basis in high volume. The CORONA system was a real workhorse. I was really excited when I found out about all the photogrammetry and mapping we were doing and saw the different kinds of capabilities that existed. It was a wonderful opportunity for me.

However, I think it's important to note that the CORONA system wasn't initially designed for mapping purposes. It was initially there to "peer behind the curtain" and give us a sense of the Soviets' military might. Nevertheless, it quickly evolved into supporting mapping capabilities.

There were many complexities to the mapping process that we hadn't dealt with before CORONA. There were horizon and stellar cameras, and other technologies that could help us fix the location of a satellite so that we could get good geodetic positions of points on the ground. Every day, intelligence people were coming in with new discoveries—new missile sites, new complexes, even new cities.

We were mapping much of the Eurasian landmass on a daily basis in the mid-1960s. Requests for detailed information and intelligence came in to us regularly. People would ask us questions like, "Exactly where is this site located? What are its exact dimensions? Which way is this missile pointing?" We also did a lot of "line-of-sight" types of estimates at NPIC, which answered the question, "If I'm standing here, can I have a direct line-of-sight to another spot in this location?" In short, there were all sorts of detailed questions that challenged us to add value and dimension to what we were actually seeing on the film.

It is also important to remember the context in which we had to deal with things in those days. For example, we didn't have PCs [personal computers] or any other type of individual computer support. We did all of our centralized photogrammetric jobs on a mainframe instead. We'd get two or three jobs processed a day. We did most of them overnight. They were processed on punch cards. We also generated a lot of paper tape that we would then feed into plotters to get more detailed site maps. And we didn't have handheld calculators in 1965; during that period, we had to look up trigonometry functions and use slide rules. Furthermore, as often happens, the ground resources lagged behind the overhead satellite system, so that when the satellite systems were coming into their own, there were a lot of folks on the ground who were doing the real yeoman's work with new tools that were just evolving. They had to use those tools to get the full range of information that was available from the imagery that was coming through the door.

William C. Mahoney, another CORONA photogrammetrist, began his career as an Army topographical officer during the 1940s and worked his way up to a position as a photogrammetrist at the Air Force's Aeronautical Chart Information Center (ACIC) by the early 1960s.[2] Mahoney recounts—in even greater detail than Gifford—the challenges mappers faced as they developed a global mapping system. He also elaborates about the processes and techniques mapmakers used to exploit CORONA's information to its full potential.

Imagery from a 1963 CORONA mission. A number of Hen Roost and Hen House phased-array radars located at the Sary Shagan antiballistic missile complex along Lake Balkhash are visible at the center of the image. Sary Shagan was a high-priority target for U.S. intelligence. Similar radar facilities were located throughout the Soviet Union, but not in the concentration seen at Sary Shagan.

Before I went to work for the ACIC, I was studying analytical triangulation at Ohio State University. Up to that time, anybody who was interested in that field had been forced to make monoscopic measurements on photographs and then manually compute and record the complex records of what he or she measured. The process took a long time. The most advanced computers available at that time (1957) were severely limited. What we were doing at Ohio State was making analytical triangulation practical by using a first-generation IBM 650 digital computer. To do so, we seized upon an idea proposed by Dr. Paul Herget—which was based upon vector mathematics—to analytically extend a strip of photographs, similar to the model-by-model manner used to extend photogrammetric triangulation strips using an analog optical system. His concept did not require a high-capacity computer solving large arrays of least squares equations. To satisfy part of the requirements for my Ph.D. degree, I developed and demonstrated the first successful analytical triangulation system by applying Dr. Herget's concepts and running a 400-mile frame photogrammetric strip over the United States using reconnaissance-type photographic materials.

I accomplished this by basing the system on stereo-comparator measurements rather than the traditional mono-comparator photo measurements that photogrammetrists had relied on in the past. When I first started to apply the Herget method, the two-plate stereo comparator didn't exist. We developed our own by converting a WILD A7, a first-order stereo-optical plotting instrument, into a stereo-comparator. This allowed us to bypass standard distortion-free mapping photographs, which had previously limited our use of such instruments.

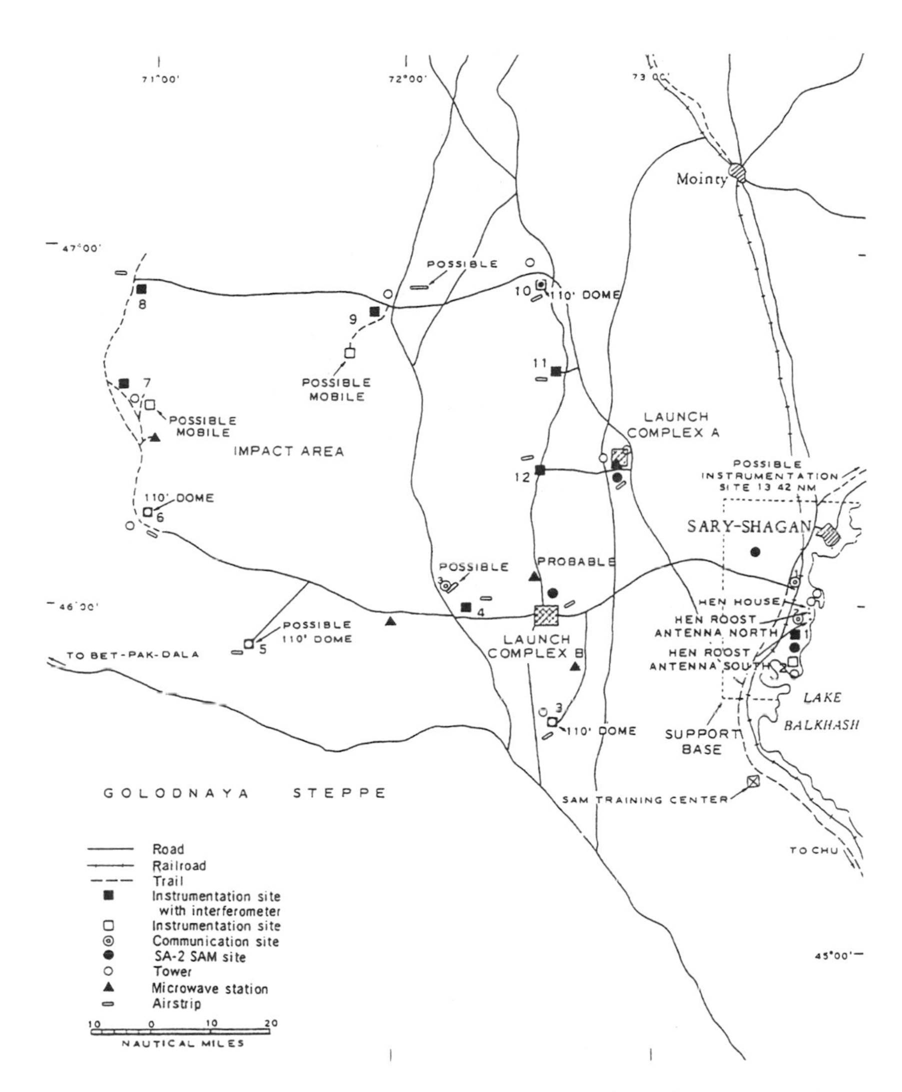

A map of the Sary Shagan antiballistic missile complex developed from CORONA imagery. A vast array of advanced tracking radars was located on the shore of Lake Balkhash and used to track Soviet reentry vehicles fired at Sary Shagan during tests of ABM systems.

In June 1959, I was recruited to the ACIC (Saint Louis) by Tom Finney, who was the head civilian of the center. He had kept track of my work at Ohio State, and, as soon as he saw that I was getting ready to leave OSU, he approached me. The interesting part was that three days after I joined the ACIC, all of my paperwork had been completed so that I was fully cleared into every photo-intelligence collection system that existed at the time. Finney had done all of the preliminary work so that there would not be any time wasted getting me on board.

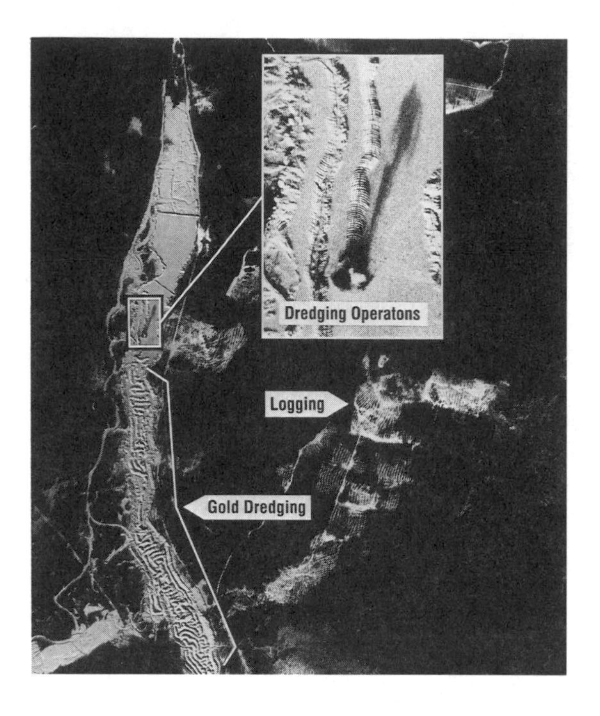

Gold dredging and logging operation in Siberia, November 26, 1970. Such photographs now allow environmental scientists to gauge the extent of environmental degradation within the former Soviet Union. (Photo courtesy National Photographic Interpretation Center)

The first challenge I faced at the ACIC was to get my OSU analytical triangulation system up and running and start using analytical photogrammetry to handle U-2 photographs. The main task was to extend photogrammetric control from areas of known ground control to position targets in the Soviet Union. Previous technologies had dealt with 6-inch, calibrated frame photographs taken while flying regular flight patterns. While it was obvious that the U-2 system had some useful mapping applications, it still presented photogrammetrists with challenges that they had never faced, including an extremely long focal length, a large frame format, uncalibrated images which covered a wide cross-field angle with no timing or attitude information, vertical to extreme tilt angles, and erratic flight lines which turned as much as 90 degrees or more within a few frames and covered very little or no ground control (either relative or absolute).

As a result of our work with the U-2, we concluded that the MC&G (Mapping, Charting and Geodesy) community, with its equipment and training, was totally inadequate for handling extremely high-resolution photographs for mapping and target positioning purposes—I mean totally! But our work at the ACIC did provide us with a wonderful opportunity to apply our imaginations and our desires to experiment with new theories, techniques, and equipment so that we could exploit these materials for mapping.

What was the MC&G's mission? We were committed to closing the U.S./Soviet missile gap. We were using the new reconnaissance systems—the U-2 and CORONA—to detect and catalog missile targets. However, we had very little geodetic knowledge to tie these targets to a worldwide geodetic system for ICBM targeting.

Our information consisted of a series of loosely coordinated Russian geodetic surveys for which many of the control stations were not photo-identifiable. Our Soviet bloc map collection was inadequate. Existing MC&G source data limited our ability to pinpoint Sino-Russian targets with more than a two- to three-mile degree of accuracy. It was even possible to make errors of up to 30 miles in some regions of Russia. In order for the MC&G to live up to its responsibility of helping to close the missile gap, we had to overcome these deficiencies with enough precision so that we could target our ICBMs against the Soviet sites.

In order to hurl an ICBM from a launch point in the United States to a target in the Soviet Union, three physical conditions would have to be met: (1) we would have to achieve an exact knowledge of the three dimensional geometric shape of the world "geoid"[3] by referring to a common datum encompassing the launch and the target points; (2) if a vehicle was launched, we would have to account for the way gravity influenced its trajectory throughout its entire flight; and (3) we would have to determine the precise geodetic position of both the launch site and the target.

The world geodetic community had already started to address the first of these three conditions by tying together worldwide geodetic surveys, both on the ground and from space. It was only possible to tie the system together from space because of CORONA. We needed a photographic collection system that was geared to meet MC&G requirements. To get it, we had to "piggyback" on the Intelligence Community without compromising the intelligence collection function. We had to upgrade CORONA and augment it with precise timing, scan calibration, and attitude readout in order to enable the MC&G's processes to control the geometry and distortions of pan geometry. MC&G had to develop a whole new analytical photogrammetry technology from scratch. It needed sophisticated mathematical models, computer power, measuring equipment, rectifiers, enlargers, and printing devices compatible with CORONA's high resolution and geometry. Old tri-met templates and inadequate optical stereo plotters were all the ACIC had to start with. We had to design and develop exploitation equipment and computer software for production application. Finally, and most important, we needed a trained workforce in order to build it. The workforce that we had in our plants, and I am counting the Army, Navy and Air Force, was using World War II technology. We had to wean them off of existing Multiplex/Kelsh conventional mapping techniques and train them to handle the new technologies based on CORONA's collection data. The development of MC&G capabilities both instigated and followed the evolution of the CORONA, ARGON, and LANYARD systems.

The first mission we received material from was the KH-1 CORONA. When the materials arrived, we were absolutely amazed by their content and detail. We said, "Boy! This is going to be great material for compiling and revising 1:250,000 maps and 1:200,000 ATCs [Air Target Charts]." Preliminary experiments revealed that the KH-1's pan geometry was manageable, but, as expected, was totally incompatible with the exisiting state-of-the-art stereo analog mapping equipment. Nevertheless, its imagery content and stable geometry suggested its great potential for mapping and targeting using analytical photogrammetry technology.

The CORONA managers and contractors' attitudes and actions were also an important factor in bringing about major changes in mapmaking during the period. One day, for instance, we were talking with some Itek representatives, and we mentioned to Ron Ondrejka, "You know, Ron, if we could just get a framed camera in this thing to calibrate the panoramic imagery, we'd have a first-class mapping system." So we talked and talked, and about a month later, Ron came back and said, "You know, I think we can get that camera you want because we've found a hole in the CORONA vehicle where we can pack it." Believe it or not, by the time the KH-4 system came along, we had our little 1½-inch camera system sitting inside that vehicle. It didn't have all the bells and whistles that we would've liked to have on it, but the KH-4 "bird," as we called it, was the first bird that was adapted to support MC&G mapping and target location requirements. Further improvements included adding a more precise timing system for frame exposure, and horizon and stellar cameras for attitude determination.[4] With these additions, CORONA started to become a true geodetic camera system.

By the time the KH-4A imagery began rolling in, we were able to make accurate panoramic photogrammetric measurements. Pan photo rectification had been established. Orbit positions were more accurate and the time of frame exposures was becoming more precise, which facilitated the beginning of analytical triangulation techniques using both orbital and attitude constraints. MC&G was finally in a position to meet the ATC requirement of 300 feet (linear error—90 percent) for relative vertical accuracy anywhere in the Soviet bloc. We were also able to use the new material to do stereo-compilation and point positioning. The KH-4A and B added other improvements. In addition to improving the on-board cameras, Doppler was added for improved orbit determination. MC&G multiple orbit strip and block triangulation adjustment techniques were developed to better calculate the "drag" variable affecting change in the movement of the space vehicle. These adjustments were also designed to improve the consistency of joins between frame models over large areas of the Soviet bloc. The WGS 66 [World Geodetic System, 1966] ellipsoid was constantly improved in order to achieve the WGS 72, which better defined the earth's geoid and gravity field.[5] These accuracies determined the way the Minuteman system evolved by allowing the use of a single missile to carry multiple lower-yield warheads.

CORONA wasn't just an intelligence and target location tool. It was also a major instrument for compiling the cartographic and digital products that were necessary to support ground forces. By 1969, we could reliably predict any position on the earth's surface to within 450 feet with a 90 percent accuracy rate. That rate, besides meeting the SAC's ICBM target location requirements, also met the DMA's [Defense Mapping Agency] mapping requirements for medium-scale maps. CORONA also got into the business of supporting air tactical weapon systems. During the Vietnam War, for instance, we took CORONA stereo pan image models, geodetically calibrated them, and passed them out in the field as point-position databases (PPDB). New reconnaissance aircraft photographs were then correlated to these PPDBs. As a result, field forces could get a quick response target position readout over densely vegetated areas within two to four hours. Photoreconnaissance materials from the SR-71 Blackbird—which was being flown over Vietnam at that time—would come in. Targets would be identi-

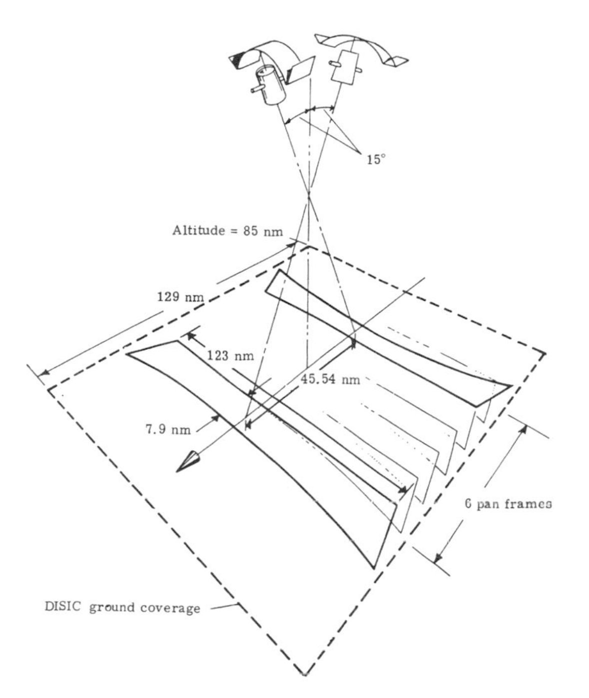

A comparison of the Dual Image Stellar Index Camera (DISIC) mapping camera coverage and the stereoscopic panoramic camera coverage. The Index and later DISIC cameras were used both for mapping and for locating the position of target objects on the ground.

fied and located by correlating them to pan PPDBs, to give target coordinates for future weapon strikes. This approach to ground force targeting addressed the major difficulty in Vietnam of finding targets under heavy tree cover. The only way you could determine a precise target location was by doing a stereo correlation to the pan PPDBs. Another application was the use of CORONA pan materials to compile TERCOM (terrain contour matrices) digital terrain matrices, which were used for the in-flight guidance control of cruise missiles.

By WGS 72, MC&G had developed an accurate mathematical geoid and gravity model of the earth. Various geodetic datums from around the world were tied together using this model to create a world geodetic system. This information, in turn, was fed into CORONA's guidance and control to improve its orbit. Improved orbit data was then fed synergistically back to MC&G to improve its target location capability.

To complete the story of the impact of CORONA, I again call your attention to the fact that before CORONA, DoD map compilation was based on the use of standardized 6-inch focal-length frame, distortion-free, low-resolution stereophotographs processed in Multiplex/Kelsh and other commercially available analog instruments. To make the transition to CORONA, a major technological revolution also had to be accomplished in the management and growth of DoD MC&G mapping organizations, including the in-house training of the DoD MC&G's workforce to use analytical techniques; the redesign of the DoD MC&G's production process; and the coordination of DoD MC&G's research and development through the Army, Navy, and Air Force labs.[6]

The techniques offices were associated with production elements so that every element essentially had its own R&D team that could solve its production problems. Originally, there was a conflict over whether there should be one techniques office solving everybody's problems, or whether several techniques offices should head up smaller production teams. The site-centered management method brought immediate improvement.

In summary, the impact of the CORONA system on MC&G was profound. It got us into the satellite business and made us develop new techniques and new instruments that enabled us to handle panoramic material as well as any type of future image-collection system. Eventually, as a result of the entire MC&G community's expansion, an OMB Federal Mapping Task Force was formed in 1972, and a study was conducted which resulted in all DoD MC&G resources being consolidated under the Defense Mapping Agency.[7]

Kenneth I. Daugherty was one of Mahoney's colleagues at the ACIC. Unlike Mahoney, Daugherty was a geodesist, not a photogrammetrist.[8] While geodesists and photogrammetrists often worked together very closely during the CORONA program, there was still an important distinction between them. For example, geodesists attempted to map the world based on ground surveys, whereas photogrammetrists mapped the earth based on aerial photographs. Speaking from a geodesist's perspective, Daugherty details the training process that such specialists went through during the CORONA era and how they developed a global mapping system based on a series of old ground surveys plus CORONA images. He also explains how mapmakers would then take suspected enemy targets and locate them on the new map system and gain a more complete intelligence picture. Recounting the complexities of the CORONA mapping process from the very beginning, Daugherty notes:

There were only a few courses in geodesy offered in this country at that time. Syracuse University was one of the few schools that offered geodesy at the undergraduate level. Geodesy was not taught at the graduate level in this country until Ohio State formed a graduate program in the 1950s. Geodesists were traditionally trained on the job or in Europe, which was the center of geodetic thought and practice. With financial support and encouragement from the Department of Defense and the CIA, however, Ohio State imported, almost lock, stock, and barrel, a European faculty that established a program at the M.S./Ph.D. level in geodesy, photogrammetry, and cartography in the early 1950s. Out of that program came folks like Bill Kola, Bill Mahoney, me, and many others.

During the 1950s, surveys were based on local datums. Two such datums were the North American Datum of 1927 and the European Datum of 1950, which was the result of an adjustment made by the Army Map Service after World War II. We also had datums of the Far East which had been tenuously calculated by using long lines of triangulation and some measurements across the ocean. Asia was tied to Europe by a survey along the trans-Siberian railway and then an arc that went down through India.

The Roman Ruins at Lejjun, Jordan, September 29, 1971. Note the heavy defensive walls. CORONA revealed many terrain features not visible by other means. (Photo courtesy National Photographic Interpretation Center)

During this period, we had some rather forward-looking folks in the Air Force who went out and started to build a program by hiring folks who were already in the business, like Bill Mahoney. Then those people would hire folks who were just coming out of college, like me. I had a brand-new degree in mathematics, geography, and geology and was recruited by the ACIC. The recruiter said to me, "We're really interested in your cartography and your geography skills, but your math probably won't hurt either." So, within a year I was doing projections and grids on a desk calculator at the ACIC. Then they came around to me and said, "How would you like to go to Ohio State and study geodesy?" And I said, "All right!" Then I looked in the dictionary to see what the hell "geodesy" was.[9]

A large group of us, 45 civilians from ACIC and five officers from SAC, went to Ohio State, and it was one of the most intensive educational experiences I had ever had. The folks at OSU followed the ACIC's advice and made sure that since we were getting paid eight hours a day, that we remained in the classroom eight hours each day.[10]

After training at OSU, we came back to the ACIC and started to work on putting together a geodetic system in which you could plot points from different continents on a coordinate system. This would allow us to deploy ballistic missiles accurately on target from one continent to another, and to track satellites. So we came up with a World Geodetics System in 1959, 1960, 1966, and 1972. Every day, we were doing new and revolutionary work because there were no reference books on the subject, and there were no programs or manuals. We had to invent everything from the outset.

Our source materials on the Soviet Union were pre-Revolution 1917 surveys and early surveys of the communist era that had been captured by the Germans during

World War II. We used methods like looking along the trans-Siberian railway for the old wooden triangulation towers that the Russians had used when they did some of their previous surveys. When CORONA flew over, we would go and find these towers—which were still sitting there rotting along the trans-Siberian railway—and correlate their geodetic coordinates, which we already knew from the old surveys, with the new satellite photos we had obtained.

We also soon realized that what we needed to do was survey the entire earth from space. We could not correlate the mishmash of information about an area without such a survey. We needed a common denominator to hold it together. So we came up with a process that required a satellite with a calibrated camera, a precise orbital position, and an exact camera attitude. As we fed each of these factors into the system, they were incorporated into the CORONA program and into later programs to the point where we could survey from space and create accurate maps and targeting materials. After all, east of the Urals, we only knew where some places were within 15 to 30 miles, and some places we didn't know about at all.

In 1966 or 1967, mostly out of intellectual curiosity, I also decided that since we were doing what we called G&G calculations (i.e., the geodetic and gravity effect on ballistic missiles), I would sit down and do some reverse calculations by making assumptions about what the Soviets knew based on the information that was available to them. In fact, they could go out and buy the USGS (U.S. Geologic Survey) quads which unintentionally revealed our missile fields by showing where our missile fields' fences were located.[11] My reverse calculations uncovered that they did not have the same problem we had. The targeting problem was much simpler for them. They could go out and, for less than one dollar, buy a 1:24,000 map and then actually go out and walk around a missile field's fence. So, from the outset, their ability to have a more precisely targeted system was there. Whether they did or not, I can't say for sure.

During this period, we were also working on a program to improve the one-nautical-mile accuracy of the Atlas ICBM system to something that would be closer to a tenth of a nautical mile in the Minuteman series. For those of you who visited St. Louis during that era, you will remember that they were building the St. Louis Arch. Interestingly, the St. Louis Arch is almost exactly a tenth of a nautical mile in height. What we were planning to do in the ballistic missile and geodetic world was to stand 5,400 nautical miles away from a target and lob every other missile into an area the size of that arch. This analogy gives you a visual picture of what we were trying to accomplish.

Because gravity variations affected the way satellites and ballistic missiles moved, we also set out to build a gravity library so that we could correlate gravity information from all over the world and improve the "Earth Gravity Model," which in turn helped us define the geoid. We managed to accomplish these goals in a rather rudimentary fashion for WGS 66, in a much more refined fashion in 1972, and then again in another 12 years with WGS 84. Today's accurate worldwide navigation and surveying capabilities came about as a result of WGS 84 and the Global Positioning System.[12] Back in 1967, on behalf of the community of geodesists and folk of that ilk, I briefed Dr. Harold Brown, who was then DDR&E [Director of Defense, Research, and Engineering], and convinced him that the department ought to spend about

The Kodak-developed Dalton Spray Processor, which was used to provide precision processing of high-resolution CORONA film. As CORONA materials proliferated and were used by customers such as the U.S. Geological Survey, more copies of the film were needed. (Photo courtesy of the Eastman Kodak Company)

$25,000,000 to update the geodetic and gravity models so that we could improve the Minuteman I system—and he bought it. From that point on, we never looked back, and as a result, the DoD now has the world's finest geodetic and gravity source materials and the ability to exploit those materials in support of national programs.

Lowell E. Starr, the fourth and final cartographer, was a civilian mapmaker with the U.S. Geological Survey (USGS) during the CORONA era. Starr was a key innovator in the CORONA mapmaking process and helped significantly modernize cartographic capabilities during the period.[13] In his discussion, he reviews the USGS's participation in the CORONA program and the way that the various branches of the Intelligence Community cooperated in order to map the earth's surface. He also explains the important impact that the Intelligence Community's cooperation had on improving civilian maps. Starr notes:

In the mid-1960s, there was a study that the BoB [Bureau of the Budget—now the Office of Management and Budget] and the Department of Defense commissioned to analyze the best methods for exploiting CORONA materials to the fullest. The study essentially served as the springboard for getting the civilian community involved with the program. It provided civilians with the opportunity to grasp technological innovations and use them, even though we didn't get involved until about 1965.

The USGS was a civilian organization that was very conservative and scientific. We were steeped in the culture of near-academic type research—the public dissemination of maps, and data and scientific publications—and many of us were therefore reluctant to participate in a covert operation. After all, covert data did not support the "publish or perish" world's fundamental objective because the associated information could not be referenced in a publication. Nevertheless, Dr. Thomas Nolan, who was the director of the USGS at the time, and William Radlinski of the National Mapping Division saw the value of USGS participation in the CORONA program.

In early 1965, I was reassigned to headquarters from a USGS Mapping Center located in Rolla, Missouri, to serve the final year of a three-year training program. This one-year assignment actually lasted approximately 30 years. Winston Sibert, my supervisor and chief of the Office of International Activities, came to me and asked if I might be interested in participating in a special classified program which could be of substantial benefit to the USGS National Mapping Program. Well, I didn't have the foggiest idea what was going on, but I understood the general drift of what was possible, so I said, "Yes, I'm interested." At that time, the USGS was heavily involved with NASA and the Gemini and Apollo programs. So I started working on NASA/USGS cooperative activities so that I could gain experience with space imagery and programs. This occurred during a period when I was waiting to receive my clearance, which usually required only about three months in those days. I later learned the real reason I had been assigned to those activities was so that I could learn about orbital science and a few other things that were applicable to mapping using CORONA materials.

In 1966, we formed a small team of people and proceeded to pull off a near miracle—at least by U.S. government standards. Winston Sibert formed a small team with the goal of appyling CORONA materials to USGS mapping programs. He hired a security officer, program officer, some USGS production and photolab experts, and me. We were able to establish a liaison with the CIA through Dr. Joe Baclowski, the man who deserves much of the credit for the success of the civilian community's program. We then received a series of billets—I think 40 to begin with—and a token budget. We selected a site in Reston, Virginia, and built a classified facility (Reston would later become the home of the USGS headquarters). We had a great deal of difficulty building that facility in Reston. Nevertheless, what was so phenomenal about the endeavor was that we built the facility, established a staff, and occupied the building in 13 months for only about $300,000. That would be impossible today.

To staff the new facility, the USGS management gave us the freedom to visit our field offices and conduct personal interviews with employees to determine their interest in the CORONA effort. One can imagine the potpourri of individuals we amassed. Some people were innovators, others were individuals who wanted to learn, and still others wanted to use this unique experience as an opportunity to advance in the scientific community. Well, we brought these people together and managed to begin revising USGS maps in about three or four months after opening the new classified facility.

The first CORONA materials we received were KH-3 images, which we desperately needed in order to revise the 1:250,000-scale maps of the United States. These maps, which provided full coverage of the continental United States and Alaska, were originally prepared by the Army Map Service during World War II and ultimately completed about 1950. The Aeronautical Chart and Information Center (ACIC) in St. Louis provided the procedures for compiling the maps and rectifing the CORONA materials. Bill Mahoney deserves much of the credit for the innovative developments that occurred in the technology we were working with at the time.

The National Photographic Interpretation Center (NPIC) was a great help to the USGS. They opened their doors to the USGS in many ways. They provided training, equipment and great insight into the processes required to interpret space imagery. NPIC's director, the late Art Lundahl, specifically got involved and provided surplus equipment to the USGS for mapping applications. Thus, the total USGS operation was built on a shoestring—with the cooperation of many governmental organizations.

The advantage USGS had over the military and the other segments of the Intelligence Community was that we had excellent "ground truth." The original maps were accurately compiled and fitted to a well-developed project and geoid. Thus, we obtained the best rectified imagery available to us, used "green eyeshade" cartographic processes, and revised the 1:250,000-scale series for the contiguous U.S. and a great portion of Alaska. Some images of Alaska, however, were of poor quality because of the orbital inclination of the spacecraft and the cloudy conditions that often exist in that part of the world.

In 1970/1971, another milestone occurred that stimulated the use of CORONA materials in civil programs. The Bureau of the Budget formed the Federal Mapping Task Force, which was commissioned to take a look at all mapping activities in the United States. The creation of the task force encouraged some other civilian agencies to get more deeply involved in the peaceful exploitation of CORONA materials.

Some competition consequently occurred among various groups.[14] The USGS was able to obtain modest line-item funding, which in turn accelerated the capital equipment program and helped the USGS obtain state-of-the-art hardware and software. There was also an unofficial group formed in 1971 called the Civil Applications Committee (CAC), which interfaced with COMIREX. The CAC was able to take the engineering test flights, the principal source of information for civilian organizations, and closely coordinate and prioritize them. The CAC was chaired by the USGS, and served to coordinate civilian priorities until the end of the CORONA era. It continues to function today.

As technology matured, we realized another important mapping objective could be satisfied by using CORONA materials. We concluded that it would be possible to inspect how up-to-date were the 55,000 quadrangle maps of the continental United States at 1:24,000 scale (1 inch equals 2,000 feet). The process was very simple. We projected the CORONA images over the maps to ascertain whether or not they would need revision, based on the extent of the changes shown on the satellite image. For example, a trained employee could view a roll of CORONA film and quickly annotate the related maps by using USGS criteria to decide whether or not the maps would require revision. This process was so reliable that we later just reprinted some of the inspected maps and included the statement "No Revision Required, Inspected 19__." Of course, maps which required revision would be corrected by conventional methods.

In summary, there were three major accomplishments that were realized as a result of the interaction with the CORONA program. First, the 1:250,000-scale maps of the continental U.S., and parts of Alaska, were revised once, and in some cases twice. Second, several thousand 1:24,000-scale quadrangle maps were inspected by using more than 40,000 CORONA photographs. And third, and most important, we formed a cooperative relationship with the Intelligence Community and with other government agencies, which probably would not have occurred without the CORONA program.

As Gifford, Mahoney, Daugherty, and Starr attest, CORONA had a revolutionary impact on the military and civilian mapmaking processes at a key point during the Cold War. Due to the cartographic innovations that geodesists and photogrammetrists developed, and the tremendous cooperation between the civilian, military, and industrial groups involved in the CORONA program, mapmakers were able to create a sophisticated world geodetic coordinate system that has had a profound effect on the maps that defense planners and civilians have continued to use.

However, perhaps Gifford and Mahoney's final perspectives are the best way to summarize the revolutionary influence CORONA had on mapmaking, for, as Gifford notes:

Tremendous things occurred during the CORONA program. There were wonderful people doing great things in the mapping community despite the odds. After all, we faced a very demanding customer; the Defense planning community wanted the information yesterday. So photogrammetrists and geodesists had to deal with the continual stress of providing better and more accurate information in a timely fashion. We were all racing the clock. But we never had a sense that we were doing it in isolation; there was always high priority put on all of the activities that supported the mapping and dimensional areas of the CORONA system.

And that prioritization apparently paid off, because as William Mahoney notes about the mapmakers who worked on the CORONA program:

We provided the last essential element. We put the cross hairs on the target that made our counterforce credible.

11

EXPLOITING CORONA IMAGERY

The Impact on Intelligence

American imagery and intelligence analysts' work changed drastically as a result of the CORONA program. In this chapter, four former analysts discuss the significant impact that the CORONA project had on U.S. intelligence and comment on a wide range of issues, including how CORONA began and its influence on American strategic planning. The heart of their discussion focuses on the techniques and strategies that they developed to exploit CORONA imagery so that they could provide government officials with up-to-date and accurate information. They also highlight the actual discoveries they made and challenges they faced. In all, the four analysts conclude that the CORONA program was an absolutely essential component in the successful gathering and assessment of intelligence during the Cold War era.

Robert ("Rae") M. Huffstutler, a former intelligence analyst and manager at the CIA's Directorate of Intelligence during the CORONA project, provides an explanation of why the Intelligence Community desperately needed a satellite reconnaissance system like CORONA in the early 1960s.[1] He also outlines some of the questions policymakers wanted answered once CORONA became operational. For example, Huffstutler notes:

> Starting in 1949 and going through the 1950s, the Soviets, in their denied area, had a large series of successful nuclear tests. There were long-range bombers rolling off the assembly lines in quite uncertain numbers. There were ballistic missile tests, but we did not know whether there were actually ballistic missile sites, though there were a lot of reports that there were. And the 1961 launch of Yuri Gagarin and his space ride simply heightened the anxiety held by national security policymakers and caused them to want the answers to some fairly straightforward questions, which were a lot easier to ask than they were to answer. The initial questions that they were asked were of a "yes" or "no" variety. Is there a missile gap? Is there a bomber gap, and, if so, to what extent does it

exist?[2] For the answer to these questions, we had to rely very heavily on remote sensing, because it was our only look into a denied area—a police state where travel was carefully controlled and constrained.

Before CORONA, American intelligence about the Soviet Union and other countries was very fragmented. Richard J. Kerr, a CIA analyst who worked at the Office of Current Intelligence during CORONA's early years, explains just how difficult it was for intelligence officers to make accurate assessments before the advent of CORONA imagery. He also hints at how their analytical methods changed once they started getting such information.[3] Kerr remembers those years as

an extraordinarily challenging period. You know for a good deal of time we didn't know what missile sites looked like. And we didn't know their pattern. And we didn't know what a gaseous diffusion plant or a reactor looked like. Because at the beginning of this process, it was kind of a clean slate. We knew what we knew about ourselves and we had all this information—some of it out of the SIGINT [signals intelligence] world—and we had a lot of information from refugees about plants and facilities on little 3 by 5 cards.

So it started up like a puzzle, but we didn't have the big picture of what the puzzle was. So we'd put a piece down and then we'd try to figure out whether it was the corner, the bottom, the top, and then we'd try to figure out where it fit in. And for several years, we were in that business of trying to fit together these pieces, and a lot of people put those little pieces together and developed the techniques to put them together. What the people involved were doing was trying to make sense out of a huge variety of information and photos, trying to connect those installations and connect the patterns. They were trying to make sense out of it.

We had some idea of what the rough structure was, but to my knowledge we had not accurately identified the location of any major ICBM facility prior to the availability of CORONA imagery. And, in fact, some of the things that we identified as "highly probable" turned out not to be missile sites. I think Novorosissk was one of the areas that we zeroed in on and said this has got to be one. Well, it wasn't.

As this comment suggests, once CORONA began successful operations in August 1960, it started providing analysts with "new information," which, as Rae Huffstutler recalls, allowed them to sharply revise the National Intelligence Estimate of the Soviet Union's ICBM capabilities, and to carry out more strategic planning. Huffstutler notes:

Just to give you an idea of the impact of imagery on the intelligence estimating process, I pulled three quotes from three National Intelligence Estimates [NIE] which have recently been declassified. These are all from an estimate designated "11-8" which was a series on "Soviet Capabilities for Strategic Attack." The first one was published in

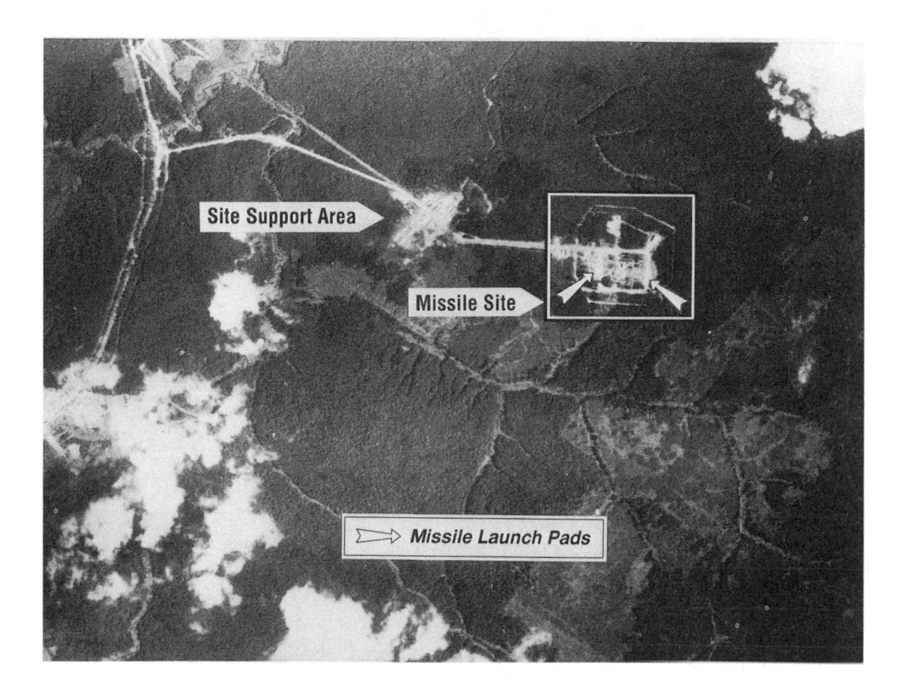

The Yurya ICBM complex, USSR, June 28, 1962. Soviet ICBM facilities were a primary target for CORONA missions. They were easily discernible by the multiple security fences that surrounded them. (Photo courtesy National Photographic Interpretation Center)

February of 1960, and in its conclusion it said that the Soviet ICBM program would probably provide on the order of 140 to 200 ICBMs on launchers in mid-1961. The basis of this estimate was the reporting of some immigrants, what HUMINT we could collect, and SIGINT. But this information was unclear about the total numbers and disposition of the forces.

In August 1960, some six months later, a conclusion was published in the next NIE 11-8, which laid out three illustrative programs using different assumptions. By this time nobody had any confidence in the facts and they tried to show that if the Soviets produced as fast as we thought they could produce, they could have 50 to 200 ICBMs on launchers by mid-1961.

However, by September 1961, there was a major change in NIE 11-8. It concluded that new information had caused a sharp downward revision in the estimate of Soviet ICBMs on launchers to between 10 and 25. In fact, I think in retrospect, we found out there were only six operational sites, plus some launchers at the test range. Obviously the difference between the patterns of evidence that you get with and without imagery is significant.

On the medium-range and the intermediate-range missiles, the fact is that we had a lot of accurate information because—being of shorter range—they were in the western

Soviet Union, and we had better access to immigrants and to reporting coming out of that area. But the impact of imagery alone makes a difference—a difference you can see when the Intelligence Community has factual data instead of speculative theories. For example, with the acquisition of hard evidence, the size of the National Intelligence Estimate went from an inch thick down to about only ten or fifteen pages. That was a particularly dramatic time in the analytical business because we could tell national security policymakers basically what the strategic balance was. In short, how bad the threat that we faced was. For the disposition, location, and size of strategic forces, there was simply no substitute for the CORONA system.

Well, where did all this lead? Once you could describe not only the strategic threat to the United States, but also the ground threat to the United States, and once you could say where the threat was positioned, you could begin to describe how large a mobilization would be necessary for the Warsaw Pact to overrun Europe. This was the basis for strategic warning and allowed senior policymakers in Washington to do strategic defense planning. There was no need to hedge against imaginary forces when we knew what we were dealing with on a factual basis, especially when we knew the size and disposition of all operational military formations.

Eventually this also allowed us to pursue strategic arms limitations. After all, it would have been impossible to establish sufficient political confidence in a verification process based only on HUMINT and SIGINT sources, especially when you always had to assess whether a HUMINT source really knew what he or she purported to know. You had to ask yourself if he or she had the whole story.

During CORONA's early years, image analysts developed a number of techniques and methods to exploit CORONA's imagery to its fullest extent. In the following section, Dino Brugioni, a senior manager at the National Photographic Interpretation Center (NPIC) during the CORONA program, details the interpretation process that analysts created and the problems they had to overcome during the program's tenure.[4]

The image analysis process is painstaking and often tedious. With the advent of multisensor technology it has grown both in complexity and sophistication. It is largely cognitive and is based on the recognition of features and patterns that are of special interest to intelligence. These special features are called signatures. For example, SA-2 surface-to-air missile sites display a Star of David pattern. An SS-5 intermediate-range ballistic missile site, in the initial stages of construction, is often referred to as "slash marks." All of these signatures are carefully cataloged and often referred to as "P.I. [photo-interpreter] keys." The imagery analyst develops a unique combination of skills and knowledge. Inherent traits include attention to detail, curiosity, inquisitiveness, diligence, deductive and inductive reasoning, and above all, good eyesight.

The use of radar, infrared, and multispectral sensors during the Vietnam conflict required interpreters to become sensitized to new forms and patterns. An experienced analyst could derive an enormous amount of information from what might have seemed to be an out-of-focus, improperly developed photograph.

Arthur Lundahl, first director of the National Photographic Interpretation Center (NPIC). Lundahl built NPIC into a major center of expertise on the interpretation of overhead photography. (Photo courtesy Dino Brugioni)

Over the years, the experienced analyst has cataloged hundreds of methods to outfox the enemy. He knows that man lives by laws, rules, customs, and practices. [Arthur] Lundahl [director of NPIC during the CORONA era] frequently addressed his new interpreters with this analogy: "Scribe a 25-mile circle on a map in most areas of the world, and man is born, lives and dies within that circle. Carefully analyze the aerial photographs of that circle, and you'll determine what man eats, what he wears, his source of water, what he cooks with, where and how he's educated, his customs, how he makes a living, his religion, the home he lives in, his interaction with nature, and finally, where he's buried."

In exploiting imagery throughout the year, we became acutely aware that each day, week, month, and season presented serendipitous benefits that could be exploited. For example, the best military order of battle is obtained on Sunday morning when most of the equipment is in garrison. Capabilities of ground forces are best observed in the spring training exercises. A heavy snowfall negates all camouflaging efforts. By tracking snow-clearing operations of military facilities, one gets the idea of the importance of each building. The headquarters building is usually cleared first, followed closely by clearing the paths to the latrines. Melting snow on a roof indicates which buildings are heated and activities in the forest are best observed in the spring before the trees begin to leaf.

Security fences which prevent ground observation of an installation are also an immediate flag for the imagery analyst. The Soviet penchant for security immediately

Although atmospheric effects frequently degraded CORONA imagery, they could occasionally make the photo-interpreter's job easier, as in this photograph, where a surface-to-air missile site is revealed by snow cover. This site was imaged by CORONA Mission 1029, launched on February 2, 1966.

aided our interpretation effort. Soviet strategic installations were ringed by not one, but by several fences and these were very apparent.[5]

When the Soviets began shipping arms and munitions to other countries, and they knew we were overflying them, they built crates to transport and conceal the military hardware they were sending abroad. Our photogrammetrists could precisely measure the crates and determine what was in them. So we developed "crateology" as part of our function. We also developed "tentology," so we could count the number of tents and get an idea of the number of troops in an area. There were all kinds of these little techniques that were developed over this period.

As satellite techniques progressed, we could easily prove that our strategic forces were clearly superior to those of the Soviets, and that U.S. superiority would continue to grow. Sherman Kent, the Director of the Office of National Estimates, after seeing our tabulation of Soviet missile, bomber and submarine forces, said, "Hell, this is no longer an estimate, it's a fact book."

Overall, each day and each frame of film was a most rewarding adventure and satellite imagery became the most valuable intelligence source of physical information. We knew that our interpretation analysis had to be precise because we knew we were only a few steps away from the president, his policymakers, and Congress.

Like Brugioni, David Doyle was an imagery analyst and manager at NPIC during the CORONA era. He specialized in the interpretation of Soviet ballistic

The Steuart Building in Washington, D.C., located at the corner of Fifth and K Streets NW, served as the site of the U.S. photo-interpretation effort for the U-2 and early CORONA missions. A car dealership, petroleum company, and real estate company occupied the first three floors of the building; the Photographic Intelligence Center (later the National Photographic Interpretation Center) occupied the top four floors. Intelligence specialist Dino Brugioni is in a trench coat at right. (Photo courtesy of Dino Brugioni)

missile systems and was one of the key imagery analysts involved in the assessment of the Soviet intermediate-range ballistic missiles in Cuba during the Cuban missile crisis of 1962. Doyle was present when the first CORONA images came in for analysis in August 1960. In the following passages, he recounts the processes, techniques, and strategies analysts developed to tease information out of the imagery. He also notes the teamwork that made the CORONA program so successful.

When the film from Mission 9009 [the first successful CORONA mission] arrived at the Steuart Building at 5th and K Streets in northwest Washington, it was the start of a new era. That mission gave us more coverage than all the U-2 missions. And that was to be repeated on every day that a CORONA satellite was in orbit. So, over a year's time, it was a tremendous amount of information at our disposal.

Photo-interpreters using micro-stereoscopes in the Steuart Building. A rear-projection viewer used for CORONA imagery is on the right. The ceiling consisted of pressed seaweed, which constantly flaked and fell on the photo-interpreters and their materials. The facility was less than ideal for the rapidly expanding photographic interpretation effort. The building was used for photo-interpretation and mission planning from 1956 until the move to Building 213 at the Washington Navy Yard in January 1963. (Photo courtesy of Dino Brugioni)

That first mission, even with its 30- to 40-foot ground resolution, was good enough to actually see SAM [surface-to-air missile] sites on it, and many of them we identified. The same thing held true for the second successful mission in December of 1960. We hit paydirt on Mission 9017 in June 1961. It was on this mission that we found the first deployed ICBM sites still under construction and a number of MRBM [middle-range ballistic missile] sites west of the Urals. And, by the fall of 1961, we had established imagery signatures for five different deployed ballistic missile systems—the SS-4, -5, -6, -7, and -8. The -6, -7, and -8 were the ICBM range missiles.

By that time, we had reorganized how we approached things, and had broken up into different groups to increase efficiency at NPIC. We had what we called "search" teams, which were groups of two or three image analysts, two of which were looking at a rear-projection viewer that the film slowly went across. You had another image analyst that sat at a 40-inch light table with a stereoscope who could take photographic coordinates in *x* and *y*. We had another person, a collateral support person, who got maps, other data, and generally helped the image analysts in their chores. And we had specialist teams who knew offensive and defensive missiles, nuclear energy, and naval installations. And these same teams kept following the same targets. So, as we got multiple coverages of installations, we rapidly became as familiar with those installations as we were with our own neighborhoods.

In the spring of 1962, we had our first stereo coverage. That's when we added two cameras to the first KH-4 and that was as good as increasing resolution by a substantial amount. We were able to see that the "scratching" on the earth's surface was actually an excavation for a hardened group of silos that was being put in at a missile site. It wasn't just scarring on the ground getting ready for something else.

It was a very arduous search process to follow everything once you found out that the Soviets were deploying missiles, or doing something unexplained. It was all very time-consuming. The intensity of your analysis was the big variable. You took a difficult situation and you could spend literally weeks going over a couple frames of imagery if you were really looking for very, very subtle signatures. Other times, you could do many frames of imagery in a day if you were just looking for growths, such as new road constructions or something like that.

How did we approach the deception problem where they would have rubber submarines or rubber airplanes or models? We did come across some of those. Luckily, human beings are not totally disciplined and over time make mistakes. The rubber deflates and people are too lazy to go out and blow it back up again. The tail on the wooden airplane falls off, or a piece of equipment doesn't move for two and a half years. Or they build a missile site with no road to it.

Also, the maps that we had in the early days were really quite rudimentary and we had to update them for a large part of the Soviet Union. We found railway lines, power grids, and microwaves, and all of these things had to be cataloged and literally tens of thousands of new targets were entered into a database, which in those days wasn't simply sitting down at a word processor and putting it in and having the computer do all the work for you. It was writing out the information in longhand, and then having a data-entry clerk put it onto IBM cards and running those cards through the computer. It was a very manpower-intensive process that went on in those first years. Those readouts then were published in hard copy over the next few days and sent out electrically on the old low-speed printers as basic reports, which tied up many communications lines for hours and hours on end all over the defense establishment.

We had a lot of "possibles" and "probables" thrown around in those reports. In general you can interpret a "possible" to mean a 50/50 chance of what it might be and that depends on how the individual P.I. feels on the day he's writing it up. The "probable" may mean 90 percent, 95 percent, or if the P.I. is real conservative, 99 percent. We still use that to this day in imagery analysis reporting. It gives a feeling to the person reading the report of just how sure we are of the situation. Over the years we established not only the fact that you can find ICBM sites; we also learned more about their organization by tracing cable lines between silos. We could see that they were being deployed in groups of six or ten and then we could separate them by when they were being constructed at launch sites in launch groups. Then, after the sites went operational, we could determine that the Soviets had, depending on the system, a two- or

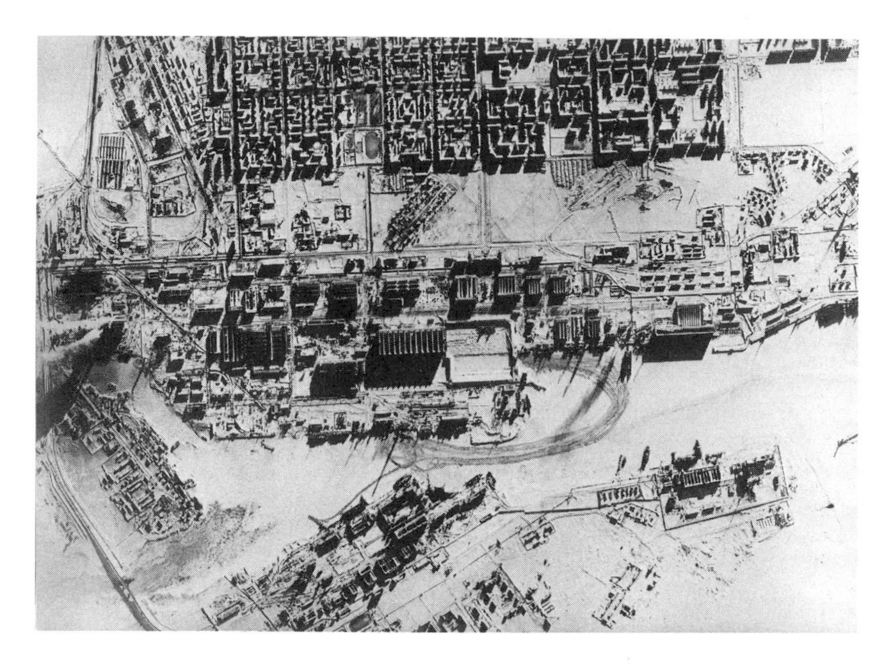

The Severodvinsk shipyard, USSR, on February 10, 1969. A large missile-firing submarine assembly and outfitting facility is located at the center of the photograph. Note the ice-breaker tracks in the river indicating that naval vessels are about to leave. (Photo courtesy National Photographic Interpretation Center)

three-year maintenance cycle where they would remove the missiles, or went in and worked on them, so that the sites were off-line for a while.

We could do the same thing for a ground forces installation. We could tell what kind of unit was there, whether it was a regiment or a division, and to some extent what their readiness was in terms of how and what they had assigned to them.

Overall, this system was a success, especially when you consider that over a 12-year period it went from 30 or 40 feet [of resolution] down to 5 or 6 feet, and increased the amount of imagery per mission ninefold. You can see that a number of improvements were made.

Despite the program's overall success, imagery analysts were still sometimes unable to answer some intelligence questions. For example, sometimes they had trouble accurately evaluating several sites and installations in a timely manner. As Richard Kerr explains:

We were not perfect in this process. We had a lot of anomalies. We had an anomalies list at one time and it was significant. There were tens of tens of tens of facilities that were big and complex and surrounded by triple barbed wire. Sometimes an airplane

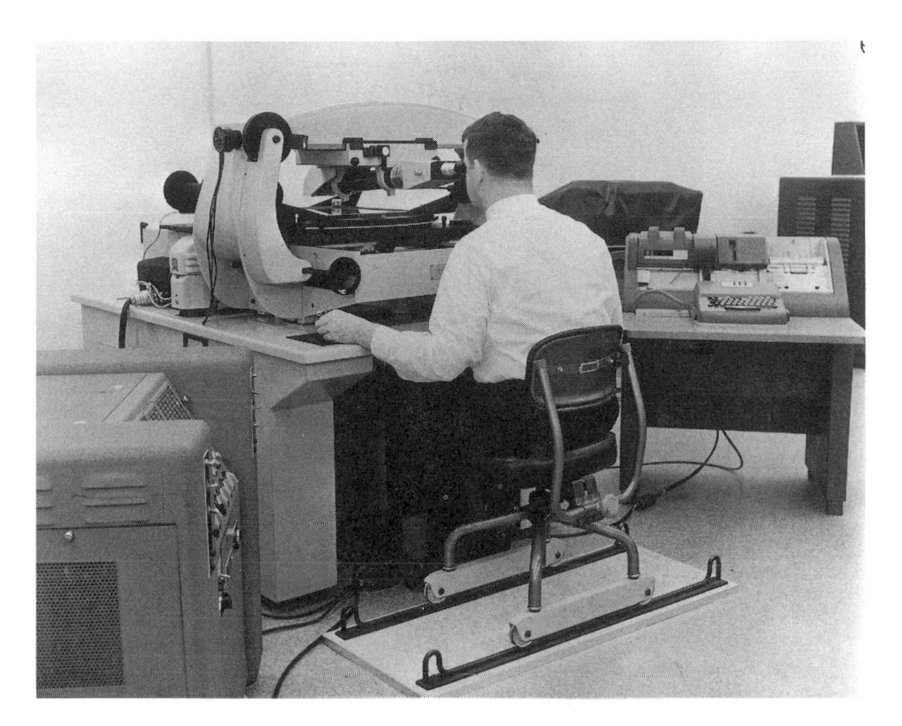

A Mann stereo-comparator used for precise measurements of strategic installations and objects. (Photo courtesy of Dino Brugioni)

would be sitting inside with its engine turned on and we'd say, "What the hell is this?!" You know, some of these things we never figured out. We called them "sensitive operation complexes."

David Doyle adds:

Sometimes analysts became overloaded and missed targets and made misidentifications. Sometimes we found just about everything, but some of it was later than it should have been. I think our batting average was pretty good, but we certainly weren't perfect. We misidentified some things and had to recant and tell them, "That's not a missile site, that's a haystack." These things happened in the process, but, over time and through repetitive coverage, we ended up getting a pretty accurate readout and feel for installations and overall capabilities.

As this last statement suggests, regardless of such miscalculations, analysts were still able to make a number of important discoveries and assessments by using CORONA imagery. As Doyle, Huffstutler, and Brugioni recount, there were many specific sites, installations, pieces of hardware, and world events that the

U.S. Intelligence Community learned about through CORONA. For example, as Doyle notes:

Every satellite pass in those days was a new event. We were seeing areas of the Soviet Union that had not been seen since World War II. We found cities that people had only heard of. We followed a lot of different things. We saw, at the Sary Shagan building, that initial ABM [antiballistic missile] prototype which was later deployed at Moscow. The patterns were the same, so we knew what we were dealing with. The Soviets were pretty well deployed out in the Far Eastern maritime provinces and in the West.

CORONA also was a critical source of information along the Sino-Soviet border. It was the best source for telling U.S. policymakers just how many divisions the Soviets had along the border and what the Chinese had on the other side. And we followed that in detail over the years. We followed their missile systems. The Shuang Cheng-Tzu missile test center was found by CORONA and we followed it extensively during China's subsequent ballistic missile deployment.

Huffstutler remembers that when the Russians started constructing the Berlin Wall in 1961, CORONA film gave

President Kennedy a very good idea of what was around the Berlin area and what kinds of reinforcements from the Warsaw Pact were available in the event they decided to either seal off Berlin, or decided somehow to take over the entire Allied sector of Berlin. That helped senior foreign policymakers to keep from making mistakes.

In a similar vein, Brugioni recalls that during the Cuban missile crisis of 1962, CORONA allowed

Dave [Doyle] to follow the intermediate-range ballistic missile sites in Russia, and make this call: "This is an intermediate-range ballistic missile site [in Cuba]." He had to convince [U.S. Attorney General] Bobby [Kennedy] and [Secretary of Defense Robert] McNamara that the patterns that we were seeing in Cuba were *exactly* those we'd seen in Russia.

And, on a more inclusive note, Brugioni reveals:

Among the world events captured on this film were the Russian and Chinese nuclear programs, and the Six-Day War between Israel and the Arab nations. We could look and see how in a matter of about three hours, the Egyptian, Jordanian, and Syrian air forces were taken out by the Israelis. You could see the Soviet invasion of Czechoslovakia; the Soviet race to the moon; the Chinese/Soviet border conflict; the India/Pakistan war; the Vietnam conflict; the Kystym nuclear incident; the construction of the Berlin Wall; the Chinese takeover of Tibet; and you can also see the remains of the Gulag Archipelago.

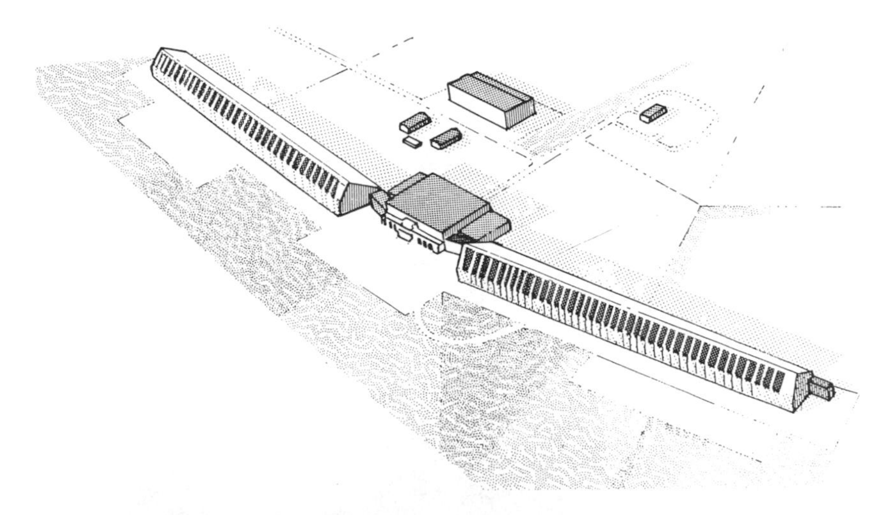

An illustration of a Hen House phased-array radar system and a CORONA photograph of a similar facility. Such line drawings were frequently produced from satellite photographs in order to conceal the technical capabilities of the satellite cameras. Hen House radars were used primarily for tracking space objects and served well into the 1990s. (Illustration courtesy Teledyne Brown Engineering)

In sum, CORONA provided analysts with the imagery necessary to make some vital assessments about other countries' military, political, social, economic, and cultural status during a difficult period in foreign relations. As Kerr, Huffstutler, and Brugioni note, CORONA has continued to have an impact on some of the problems and challenges that the United States faces, including military strategic planning and even environmental issues. Kerr, for example, thinks that

the CORONA project was quite honestly an extraordinary period. It was the business of imagery, and the business of taking that information and integrating it into a product. The problem was then to try to figure out how to control it, what to do with it, and how to put the information into context, and in some ways, that's the magic, at least from my perspective, that we did.

Are there lessons to be learned from CORONA that would apply to current threats from missiles, whether ICBMs or others? I think the simple answer is absolutely yes. I don't think there's any question that there's a transferability here of information and knowledge. This breadth and depth of knowledge and understanding of facilities and movements and activities and equipment is fundamental to being able to understand new developments in the former Soviet Union and other countries.

Rae Huffstutler adds the following assessment of CORONA's overall influence:

It was imagery that set the stage for the arms limitation talks. We began drafting the verification capabilities of the Intelligence Community in some interagency papers written in 1968 and 1969, several years before the dialogue began. With imagery, we could go to a numbers-based strategic arms limitation negotiation with the high confidence that we didn't need any help from the other side to verify it. The first phase of the arms limitation negotiations which dealt with either pulling forces back from a border, taking them out of a zone, or forbidding the presence of specific kinds of weapons systems, was only possible because of CORONA imagery. It gave us a politically acceptable way of demonstrating if forbidden forces were present, or if the treaty had been violated. That simply would not have been possible with other sources of evidence.

And Dino Brugioni notes that

the information gleaned from the CORONA program saved millions of defense dollars. Defense planners came to NPIC and, with information gleaned from satellite imagery, could plan counterweapon programs against our adversaries more effectively, more efficiently, and much cheaper.

New methods and equipment had to be designed. The microstereoscope and advanced comparators were developed to speed the interpretation effort. Some of the equipment designed to enhance satellite imagery by the National Photographic

Interpretation Center has been found by researchers to aid in the detection of breast cancer.

But perhaps the best overall assessment of CORONA's impact on intelligence is Rae Huffstutler's:

> The great contribution of intelligence during this 1960s period—and I think particularly a contribution of the CORONA system and all those that worked on it—was to help the United States plan a safe passage through a very dangerous period where a lot of mistakes could have been made had you simply assumed you understood the enemy, his intentions, and his capabilities, but had not based those judgments on hard data and the facts that came out of imagery.

Appendix A
PROGRAM OVERVIEW AND CAMERA DATA

PROGRAM OVERVIEW

	KH-1 CORONA	KH-2 CORONA	KH-3 CORONA	KH-4 CORONA
Period of operation	1959–60	1960–61	1961–62	1962–63
Number of SRVs	1	1	1	1
Mission series number	9000	9000	9000	9000
Lifetime	1 day	2–3 days	1–4 days	6–7 days
Altitude (nm)				
Perigee	103.5 (est)	136.0 (est)	117.0 (est)	114.0 (est)
Apogee	441.0 (est)	380.0 (est)	136.0 (est)	224.0 (est)
Missions				
Total	10	10	6	26
Successful	1	4	4	21

	KH-4A CORONA	KH-4B CORONA	KH-5 ARGON	KH-6 LANYARD
Period of operation	1963–69	1967–72	1962–64	1963
Number of SRVs	2	2	1	1
Mission series number	1000	1100	9000A	8000
Lifetime	4–15 days	18 days	—	—
Altitude (nm)				
Perigee	—	—	—	—
Apogee	—	—	—	—
Average operations	100.0 (est)	81.0 (est)	174.0	93.0
Missions				
Total	52	17	12	3
Successful	49	16	6	1

CAMERA DATA

	KH-1 CORONA	KH-2 CORONA	KH-3 CORONA	KH-4 CORONA
Camera model	C	C′	C‴	MURAL
Type	Mono/ panoramic	Mono/ panoramic	Mono/ panoramic reciprocating	Stereo/ panoramic reciprocating
Scan angle (deg)	70	70	70	70
Stereo angle (deg)	N/A (vertical)	N/A (vertical)	N/A (vertical)	30
Lens	Tessar f/5	Tessar f/5	Petzval f/3.5	Petzval f/3.5
Focal length (in)	24	24	24	24
Optics resolution (lines/mm)	340	340	—	480
Resolution				
Ground (ft)	40	25	12–25	10–25
Film (lines/mm)	50–100	50–100	50–100	50–100
Coverage (nm)	Unavailable	Unavailable	Unavailable	Unavailable
Film				
Types	1213	1221	4404	4404
Base	Acetate	Polyester	Polyester	Polyester
Width (in)	2.10	2.10	2.25	2.25
Image format	2.10 (est)	2.19 (est)	2.25 × 29.8	2.18 × 29.8
Load (ft)	Unavailable	Unavailable	Unavailable	Unavailable
Manufacturer				
Camera	Fairchild	Fairchild	Itek	Itek
Lens	Itek	Itek	Itek	Itek

Continued on next page

CAMERA DATA continued

	KH-4A CORONA	KH-4B CORONA	KH-5 ARGON	KH-6 LANYARD
Camera model	J-1	J-3	—	—
Type	Stereo/ panoramic reciprocating	Stereo/ panoramic Constant Rotator	Frame	Panoramic
Scan angle (deg)	70	70	N/A	22
Stereo angle (deg)	30	30	N/A	30
Lens	Petzval f/3.5	Petzval f/3.5	Unavailable	f/5
Focal length (in)	24	24	3	66
Optics resolution (lines/mm)	480	480	Unavailable	340
Resolution				
Ground (ft)	9–25	6	460	4–6 (2-ft design)
Film (lines/mm)	120	160	30	160
Coverage (nm)	10.6 × 144	8.6 × 117	—	—
Film				
Types	3404	3404, SO-380[a]	Unavailable	3404
Base	Polyester	Polyester	Polyester	Polyester
Width (in)	2.25	2.25	5	5
Image format (in)	2.18 × 29.8	2.18 × 29.8	4.5 × 4.5	4.5 × 25
Load (ft)	32,000	32,000	Unavailable	8000
Manufacturer				
Camera	Itek	Itek	Fairchild	Itek
Lens	Itek	Itek	—	Itek

[a]Also available: SO-242, SO-230, SO-205, SO-121, and SO-180.

ADDITIONAL CAMERAS ON THE KH-4, KH-4A, AND KH-4B

Camera model	Index and Stellar/Index	DISIC
Type	Frame	Frame
Focal length (in)	1.5	3
Resolution		
Ground (ft)	400–500 (est)	100–400 (est)
Film (lines/mm)	—	—
Coverage	166 × 166 miles (est)	140 × 140 miles (est)

Appendix B
LAUNCH LISTINGS

PREPARED BY JONATHAN McDOWELL

No.[a]	Spacecraft[b]	Program[c]	Camera	Mass (kg)[d]	Launch date[e]	Time (GMT)	Mission no.[f]	Recovery attempt[g]	Result
1	Agena A No. 1022	R\D	—	728	1959 Feb 28	2149	—	—	Agena failed to orbit?
2	Agena A No. 1018	R\D	—	784	1959 Apr 13	2119	R\D	Apr 14	SRV lost in Spitzbergen
3	Agena A No. 1020	R\D	—	843	1959 Jun 3	2010	R\D	—	Agena fell in Pacific
4	Agena A No. 1023	KH-1	C-1	860	1959 Jun 25	2249	9001	—	Agena fell in Pacific
5	Agena A No. 1029	KH-1	C-2	870	1959 Aug 13	1900	9002	Aug 14	SRV shot to higher orbit
6	Agena A No. 1028	KH-1	C-3	862	1959 Aug 19	1925	9003	Aug 20	SRV lost in reentry
7	Agena A No. 1051	KH-1	C-4	920	1959 Nov 7	2029	9004	Nov 8	SRV failed to separate
8	Agena A No. 1050	KH-1	C-5	835	1959 Nov 20	1925	9005	Nov 21	Agena overburn, SRV sank
9	Agena A No. 1052	KH-1	C-6		1960 Feb 4	1852	9006	—	Agena fell in Pacific
10	Agena A No. 1054	KH-1	C-7		1960 Feb 19	2015	9007	—	Thor destroyed
11	Agena A No. 1055	KH-1	C-8	781	1960 Apr 15	2030	9008	Apr 16	SRV destroyed in reentry
12	Agena A No. 1053	R\D	—		1960 Jun 29	2201	R\D	—	Agena fell in Pacific
13	Agena A No. 1057	R\D	—	772	1960 Aug 10	2038	R\D	Aug 11	SRV recovered from Pacific
14	Agena A No. 1056	KH-1	C-9	810	1960 Aug 18	1957	9009	Aug 19	SRV midair recovery
15	Agena A No. 1058	KH-1	C-10	810	1960 Sep 13	2214	9010	Sep 15	SRV sank in Pacific
16	Agena B No. 1061	KH-2	C′-1		1960 Oct 26	2026	9011	—	Thor Agena fell in Pacific
17	Agena B No. 1062	KH-2	C′-2	997	1960 Nov 12	2043	9012	Nov 14	SRV midair, no pictures
18	Agena B No. 1103	KH-2	C′-3	1043	1960 Dec 7	2020	9013	Dec 10	SRV midair recovery
19	Agena B No. 1101	R\D	—	946	1960 Dec 20	2032	RM-1	—	Agena tumbled
20	Agena B No. 1104	KH-5	A-1	1048	1961 Feb 17	2025	9014A	Feb 21	SRV sank Indian Ocean?
21	Agena B No. 1102	R\D	—		1961 Feb 18	2258	RM-2	—	Midas test
22	Agena B No. 1105	KH-2	C′-4		1961 Mar 30	2034	9015	—	Agena fell in Pacific
23	Agena B No. 1106	KH-5	A-2	1061	1961 Apr 8	1921	9016A	Apr 10	SRV shot to higher orbit
24	Agena B No. 1108	KH-5	A-3		1961 Jun 8	2116	9018A	—	Agena fell in Pacific
25	Agena B No. 1107	KH-2	C′-5	1124	1961 Jun 16	2302	9017	Jun 19	SRV recovered from Pacific
26	Agena B No. 1109	KH-2	C′-6	1125	1961 Jul 7	2329	9019	Jul 10	SRV midair recovery
27	Agena B No. 1110	KH-5	A-4		1961 Jul 21	2235	9020A	—	Thor exploded
28	Agena B No. 1111	KH-2	C′-7		1961 Aug 4	0001	9021	—	Agena fell in Pacific
29	Agena B No. 1112	KH-3	C‴-1	1136	1961 Aug 30	2000	9023	Sep 1	SRV recovered from Pacific
30	Agena B No. 1113	KH-3	C‴-2	1134	1961 Sep 12	1959	9022	Sep 14	SRV midair recovery
31	Agena B No. 1114	KH-3	C‴-3	1163	1961 Sep 17	2100	9024	Sep 19	SRV failed to separate
32	Agena B No. 1115	KH-3	C‴-4	1156	1961 Oct 13	1922	9025	Oct 14	SRV midair recovery
33	Agena B No. 1116	KH-2	C′-8		1961 Oct 23	1923	9026	—	Agena fell in Pacific
34	Agena B No. 1117	KH-2	C′-9	1173	1961 Nov 5	2000	9027	—	Agena overburn, unstable
35	Agena B No. 1118	KH-2	C′-10	1175	1961 Nov 15	2123	9028	Nov 17	SRV midair recovery
36	Agena B No. 1119	KH-3	C‴-5	1309	1961 Dec 12	2040	9029	Dec 16	SRV recovered from Pacific
37	Agena B No. 1120	KH-4	C‴-6		1962 Jan 13	2141	9030	—	Agena fell in Pacific
38	Agena B No. 1123	KH-4	M-1	1274	1962 Feb 27	1939	9031	Mar 3	SRV midair recovery
39	Agena B No. 1124	KH-4	M-2	1234	1962 Apr 18	0054	9032	Apr 20	SRV midair recovery
40	Agena B No. 1125	KH-4	M-3	1225	1962 Apr 29	0030	9033	May 5	SRV sank in Pacific
41	Agena B No. 1126	KH-5	A-5	1102	1962 May 15	1936	9034A	May 19	SRV midair recovery
42	Agena B No. 1128	KH-4	M-4	1261	1962 May 30	0100	9035	Jun 2	SRV midair recovery
43	Agena B No. 1127	KH-4	M-5	1248	1962 Jun 2	0031	9036	Jun 5	SRV sank in Pacific
44	Agena B No. 1129	KH-4	M-6	1261	1962 Jun 23	0030	9037	Jun 26	SRV midair recovery
45	Agena D No. 1151	KH-4	M-7	1280	1962 Jun 28	0109	9038	Jul 2	SRV midair recovery
46	Agena B No. 1130	KH-4	M-8	1266	1962 Jul 21	0056	9039	Jul 23	SRV midair recovery

No.[a]	Launch vehicle[h]	Site	Complex[i]		Pad no.	Operation[j]	Orbit[k]	International designation[l]
1	Thor DM-18A	No. 163	VAFB	75-3	4	1003 FLYING YANKEE	(233 × 898 × 90)?	1959 Beta[m]
2	Thor DM-18A	No. 170	VAFB	75-3	4	1004 EARLY TIME	253 × 346 × 90.4	1959 Gamma
3	Thor DM-18A	No. 174	VAFB	75-3	4	1007 GOLD DUKE	—	—
4	Thor DM-18A	No. 179	VAFB	75-3	5	1010 LONG ROAD	—	—
5	Thor DM-18A	No. 192	VAFB	75-3	4	1012 FLY HIGH	220 × 734 × 80.0	1959 Epsilon
6	Thor DM-18A	No. 200	VAFB	75-3	5	1013 HURRY UP	216 × 895 × 84.0	1959 Zeta
7	Thor DM-18A	No. 206	VAFB	75-3	4	1017 CARGO NET	161 × 845 × 81.6	1959 Kappa
8	Thor DM-18A	No. 212	VAFB	75-3	5	1021 LIVID LADY	193 × 1700 × 80.7	1959 Lambda
9	Thor DM-18A	No. 218	VAFB	75-3	4	1026 HUNGRY EYE	—	—
10	Thor DM-18A	No. 223	VAFB	75-3	5	1027 DERBY DAY	—	—
11	Thor DM-18A	No. 234	VAFB	75-3	5	1029 RAM HORN	177 × 611 × 80.1	1960 Delta
12	Thor DM-18A	No. 160	VAFB	75-3	4	1030 RED GARTER	—	—
13	Thor DM-18A	No. 231	VAFB	75-3	5	1035 FROGGY BOTTOM	256 × 703 × 82.9	1960 Theta
14	Thor DM-18A	No. 237	VAFB	75-3	4	1036 LIMBER LEG	187 × 806 × 79.8	1960 Kappa
15	Thor DM-18A	No. 246	VAFB	75-3	5	1038 COFFEE CALL	201 × 763 × 80.9	1960 Mu
16	Thor DM-21	No. 253	VAFB	75-3	4	1041 SOUP SPOON	—	—
17	Thor DM-21	No. 297	VAFB	75-3	5	1046 BOXING GLOVE	187 × 991 × 81.7	1960 Omicron
18	Thor DM-21	No. 296	VAFB	75-3	4	1047 POWER TRACTOR	246 × 703 × 81.5	1960 Sigma
19	Thor DM-21	No. 258	VAFB	75-3	5	1049 TEE BIRD	193 × 636 × 83.4	1960 Tau[m]
20	Thor DM-21	No. 298	VAFB	75-3	4	1052 SPIRIT LEVEL	325 × 767 × 80.9	1961 Epsilon
21	Thor DM-21	No. 261	VAFB	75-3	5	1053 BENCH WARRANT	256 × 1062 × 80.7	1961 Zeta[m]
22	Thor DM-21	No. 300	VAFB	75-3	4	1054 FEATHER CUT	—	—
23	Thor DM-21	No. 307	VAFB	75-3	5	1055 RUNNING BOARD	301 × 652 × 82.3	1961 Lambda
24	Thor DM-21	No. 302	VAFB	75-3	4	1059 ISLAND QUEEN	—	—
25	Thor DM-21	No. 303	VAFB	75-1	1	1060 MARKED CARD	226 × 413 × 82.1	1961 xi
26	Thor DM-21	No. 308	VAFB	75-3	5	1064 HIGH WING	235 × 811 × 82.9	1961 Pi
27	Thor DM-21	No. 322	VAFB	75-3	4	1066 STACKED DECK	—	—
28	Thor DM-21	No. 309	VAFB	75-1	1	1067 CRISP BACON	—	—
29	Thor DM-21	No. 323	VAFB	75-3	4	1069 FULL BLOWER	157 × 559 × 82.1	1961 Psi
30	Thor DM-21	No. 310	VAFB	75-3	5	1070 TWISTED BRAIDS	241 × 554 × 82.7	1961 Omega
31	Thor DM-21	No. 324	VAFB	75-1	1	1073 CANE POLE	242 × 411 × 82.7	1961 Alpha Beta
32	Thor DM-21	No. 328	VAFB	75-3	4	1075 CAP PISTOL	231 × 404 × 81.7	1961 Alpha Gamma
33	Thor DM-21	No. 329	VAFB	75-3	5	1079 DEAD HEAT	—	—
34	Thor DM-21	No. 330	VAFB	75-1	1	1080 FOG CUTTER	244 × 1004 × 82.5	1961 Alpha Epsilon
35	Thor DM-21	No. 326	VAFB	75-3	4	1081 CAT FIGHT	245 × 310 × 81.6	1961 Alpha Zeta
36	Thor DM-21	No. 325	VAFB	75-3	4	1088 SILVER STRIP	241 × 413 × 81.2	1961 Alpha Kappa
37	Thor DM-21	No. 327	VAFB	75-3	4	3201 CANDY WRAPPER	—	—[m]
38	Thor DM-21	No. 241	VAFB	75-3	4	7201 CAREER GIRL	205 × 405 × 82.2	1962 Epsilon
39	Thor DM-21	No. 331	VAFB	75-3	5	7201 LONG SLICE	209 × 515 × 73.5	1962 Lambda
40	Thor DM-21	No. 333	VAFB	75-3	4	3201 TOTAL TIME	180 × 475 × 73.1	1962 Rho
41	Thor DM-21	No. 334	VAFB	75-3	5	2201 HOLE PUNCH	296 × 648 × 82.3	1962 Sigma
42	Thor DM-21	No. 336	VAFB	75-1	1	6201 LEAK PROOF	193 × 363 × 74.1	1962 Phi
43	Thor DM-21	No. 335	VAFB	75-3	4	8201 KNOTTY PINE	213 × 415 × 74.3	1962 Chi
44	Thor DM-21	No. 339	VAFB	75-3	4	5201 TIGHT SKIRT	209 × 315 × 75.1	1962 Alpha Beta
45	Thor DM-21	No. 340	VAFB	75-1	1	7201 TRIAL TRACK	211 × 722 × 76.0	1962 Alpha Gamma
46	Thor DM-21	No. 342	VAFB	75-3	5	2201 ADOBE HOME	205 × 391 × 70.3	1962 Alpha Eta

No.[a]	Spacecraft[b]	Pro-gram[c]	Camera	Mass (kg)[d]	Launch date[e]	Time (GMT)	Mission no.[f]	Recovery attempt[g]	Result
47	Agena B No. 1131	KH-4	M-9	1269	1962 Jul 28	0030	9040	Aug 1	SRV recovered, bad photos
48	Agena D No. 1152	KH-4	M-10	1260	1962 Aug 2	0017	9041	Aug 6	SRV midair recovery
49	Agena D No. 1153	KH-4	M-11	1274	1962 Aug 29	0100	9044	Sep 2	SRV midair recovery
50	Agena B No. 1132	KH-5	A-10	1134	1962 Sep 1	2039	9042A	Sep 6	SRV sank in Pacific
51	Agena B No. 1133	KH-4	M-12	1276	1962 Sep 17	2346	9043	Sep 19	SRV recovered from Pacific
52	Agena D No. 1154	KH-4	M-13	1276	1962 Sep 29	2335	9045	Oct 3	SRV midair recovery
53	Agena B No. 1134	KH-5	A-9	1177	1962 Oct 9	1835	9046A	Oct 13	SRV midair recovery
54	Agena D No. 1401	R\D	—	915	1962 Oct 26	1614	R\D	—	STARAD Radiation study
55	Agena B No. 1136	KH-4	M-14	1279	1962 Nov 5	2204	9047	Nov 10	SRV midair recovery
56	Agena B No. 1135	KH-4	M-15	1326	1962 Nov 24	2201	9048	Nov 29	SRV midair recovery
57	Agena D No. 1155	KH-4	M-16	1306	1962 Dec 4	2130	9049	Dec 6	SRV sank in Pacific
58	Agena D No. 1156	KH-4	M-17	1312	1962 Dec 14	2126	9050	Dec 17	SRV midair recovery
59	Agena D No. 1157	KH-4	M-18	1299	1963 Jan 7	2109	9051	Jan 11	SRV recovered from Pacific
60	Agena D No. 1159	KH-4	M-20		1963 Feb 28	2148	9052	—	Thor destroyed
61	Agena D No. 1164	KH-6	L-1	1598	1963 Mar 18	2113	8001	—	Agena fell in Pacific
62	Agena D No. 1160	KH-4	M-19	1305	1963 Apr 1	2301	9053	Apr 4	SRV midair recovery
63	Agena D No. 1411	KH-5	A-12	1142	1963 Apr 26	2013	9055A	—	Agena fell in Pacific
64	Agena D No. 1165	KH-6	L-2	1615	1963 May 18	2221	8002	May 20	SRV recovered from Pacific
65	Agena D No. 1161	KH-4	M-21	1558	1963 Jun 12	2358	9054	Jun 14	SRV midair recovery
66	Agena D No. 1166	KH-4	M-22	1574	1963 Jun 27	0037	9056	Jun 30	SRV midair recovery
67	Agena D No. 1412	KH-4	M-23	1263	1963 Jul 19	0000	9057	Jul 23	SRV midair recovery
68	Agena D No. 1167	KH-6	L-3	1604	1963 Jul 31	0000	8003	Aug 2	SRV midair recovery
69	Agena D No. 1162	KH-4A	J-1	1605	1963 Aug 25	0030	1001-1	Aug 29	SRV-1 midair recovery
							1001-2	Unknown	SRV-2 failed to separate
70	Agena D No. 1169	KH-5	A-11	1201	1963 Aug 29	2031	9058A	Sep 1	SRV midair recovery
71	Agena D No. 1163	KH-4A	J-2	1603	1963 Sep 23	2300	1002-1	Sep 27	SRV-1 midair recovery
							1002-2	Unknown	SRV-2 failed to separate?
72	Agena D No. 1601	KH-5	A-6	1369	1963 Oct 29	2119	9059A	Nov 2	SRV midair recovery
73	Agena D No. 1171	KH-4	M-24L		1963 Nov 9	2027	9060	—	Thor destroyed
74	Agena D No. 1172	KH-4	M-25	1375	1963 Nov 27	2115	9061	Dec 3	SRV failed to separate
75	Agena D No. 1168	KH-4	M-26		1963 Dec 21	2145	9062	Dec 26	SRV midair recovery
76	Agena D No. 1174	KH-4A	J-5	1623	1964 Feb 15	2138	1004-1	Feb 18	SRV-1 midair recovery
							1004-2	Feb 22	SRV-2 midair recovery
77	Agena D No. 1175	KH-4A	J-6	1652	1964 Mar 24	2222	1003-1	—	Agena fell in Pacific
							1003-2	—	Agena fell in Pacific
78	Agena D No. 1604	KH-4A	J-8	1623	1964 Apr 27	2323	1005-1	Apr 30	SRV-1 failed to separate
							1005-2	—	SRV-2 remained attached
79	Agena D No. 1176	KH-4A	J-9	1618	1964 Jun 4	2259	1006-1	Jun 9	SRV-1 midair recovery
							1006-2	Jun 14	SRV-2 midair recovery
80	Agena D No. 1606	KH-5	A-21		1964 Jun 13	1547	9063A	Jun 19	SRV midair recovery
81	Agena D No. 1609	KH-4A	J-7	1602	1964 Jun 19	2318	1007-1	Jun 24	SRV-1 midair recovery
							1007-2	Jun 28	SRV-2 midair recovery
82	Agena D No. 1177	KH-4A	J-10	1630	1964 Jul 10	2315	1008-1	Jul 14	SRV-1 recovered
							1008-2	Jul 18	SRV-2 recovered
83	Agena D No. 1605	KH-4A	J-12	1671	1964 Aug 5	2315	1009-1	Aug 9	SRV-1 recovered
							1009-2	Aug 14	SRV-2 midair recovery

No.[a]	Launch vehicle[h]	Site	Complex[i]	Pad no.	Operation[j]	Orbit[k]	International designation[l]
47	Thor DM-21	No. 347	VAFB 75-3	4	8201 ANCHOR ROPE	206 × 416 × 71.1	1962 Alpha Theta
48	Thor DM-21	No. 344	VAFB 75-1	1	7201 APPLE GREEN	209 × 422 × 82.3	1962 Alpha Kappa
49	Thor DM-21	No. 349	VAFB 75-1	2	5201 APPLE RIND	179 × 404 × 65.2	1962 Alpha Sigma
50	Thor DM-21	No. 348	VAFB 75-3	5	8201 BEADY EYE	300 × 683 × 82.8	1962 Alpha Upsilon
51	Thor DM-21	No. 350	VAFB 75-3	4	3201 BIG FLIGHT	207 × 670 × 81.8	1962 Alpha Chi
52	Thor DM-21	No. 351	VAFB 75-1	2	4201 ARCTIC ZONE	196 × 385 × 65.4	1962 Beta Beta
53	Thor DM-21	No. 352	VAFB 75-3	4	4201 CALL BOARD	228 × 450 × 82.0	1962 Beta Epsilon
54	Thor DM-21	No. 353	VAFB 75-1	2	1201 ANCHOR BUOY	204 × 5593 × 71.3	1962 Beta Kappa[m]
55	Thor DM-21	No. 356	VAFB 75-3	4	9201 BAIL OUT	210 × 417 × 74.9	1962 Beta Omicron
56	Thor DM-21	No. 367	VAFB 75-3	4	9201 GOLDEN RUSH	212 × 344 × 65.2	1962 Beta Rho
57	Thor DM-21	No. 361	VAFB 75-1	2	8201 CALAMITY JANE	133 × 332 × 65.2	1962 Beta Sigma
58	Thor DM-21	No. 368	VAFB 75-3	5	8201 BABY DOLL	202 × 404 × 71.0	1962 Beta Phi
59	Thor DM-21	No. 369	VAFB 75-1	1	0048 CANDY KISSES	193 × 417 × 82.2	1963-02
60	Thor DSV-2C	No. 354	VAFB 75-3	5	0583 FARM COUNTRY	—	—
61	Thor DSV-2C	No. 360	VAFB 75-3	4	0627 CAMP OUT	—	—
62	Thor DM-21	No. 376	VAFB 75-3	5	0720 NICKEL STEEL	205 × 413 × 75.4	1963-07
63	Thor DM-21	No. 372	VAFB 75-1	1	1008 FALL HARVEST	—	—
64	Thor DSV-2C	No. 364	VAFB 75-3	5	0924 GATE LATCH	157 × 522 × 74.6	1963-16[m]
65	Thor DSV-2C	No. 362	VAFB 75-3	4	0954 GREEN CASTLE	197 × 445 × 81.8	1963-19
66	Thor DSV-2C	No. 381	VAFB 75-1	2	0999 CALICO MISS	209 × 409 × 81.6	1963-25
67	Thor DM-21	No. 388	VAFB 75-1	1	1266 CHILI WILLIE	206 × 400 × 82.8	1963-29
68	Thor DSV-2C	No. 382	VAFB 75-1	2	1370 BIG TALK	168 × 467 × 75.0	1963-32
69	Thor DSV-2C	No. 377	VAFB 75-3	4	1419 GHOST DANCE	184 × 436 × 75.0	1963-34
70	Thor DSV-2A	No. 394	VAFB 75-3	5	1561 PELICAN PETE	301 × 334 × 81.8	1963-35
71	Thor DSV-2C	No. 383	VAFB 75-1	2	1353 FELLOW KING	184 × 439 × 74.9	1963-37
72	Thor DSV-2C	No. 386	VAFB 75-3	4	2437 MARK DOWN	288 × 355 × 89.9	1963-42
73	Thor DSV-2A	No. 400	VAFB 75-1	2	2268 JUMP SUIT	—	—
74	Thor DSV-2A	No. 406	NMFPA LC-I	1	2260 DRY DUNE	182 × 392 × 70.0	1963-48
75	Thor DSV-2C	No. 398	VAFB 75-1	2	1388 WATER SPOUT	184 × 371 × 64.9	1963-55
76	Thor DSV-2C	No. 389	VAFB 75-3	4	3444 GARDEN PARTY	186 × 463 × 75.0	1964-08
77	Thor DSV-2C	No. 396	NMFPA LC-I	1	3467 HEALTH FARM	—	—
78	Thor DSV-2C	No. 395	VAFB 75-3	4	2921 NICE BIRD	183 × 465 × 79.9	1964-22[m]
79	Thor DSV-2C	No. 403	NMFPA LC-I	1	3483 KICK BALL	161 × 480 × 80.0	1964-27
80	Thor DSV-2C	No. 408	VAFB 75-1	2	3236 BEAGLE HOUND	365 × 370 × 115.0	1964-30
81	Thor DSV-2C	No. 410	VAFB 75-1	1	3754 GREEN DOOR	185 × 479 × 85.0	1964-32
82	Thor DSV-2C	No. 404	NMFPA LC-I	1	3491 OLD HAT	183 × 483 × 85.0	1964-37
83	Thor DSV-2C	No. 413	VAFB 75-3	4	3042 LONG LOOP	185 × 450 × 80.1	1964-43

No.[a]	Spacecraft[b]	Pro-gram[c]	Camera	Mass (kg)[d]	Launch date[e]	Time (GMT)	Mission no.[f]	Recovery attempt[g]	Result
84	Agena D No. 1603	KH-5	A-22	1325	1964 Aug 21	1545	9064A	Aug 27	SRV midair recovery
85	Agena D No. 1178	KH-4A	J-11	1606	1964 Sep 14	2253	1010-1	Sep 19	SRV-1 recovered
							1010-2	Sep 24	SRV-2 recovered
86	SS-01A No. 1170	KH-4A	J-3	1620	1964 Oct 5	2150	1011-1	Oct 9	SRV-1 recovered
							1011-2	Oct 12	SRV-2 failed to separate
87	SS-01A No. 1179	KH-4A	J-13	1667	1964 Oct 17	2202	1012-1	Oct 21	SRV-1 midair recovery
							1012-2	Oct 23	SRV-2 recovered from Pacific
88	SS-01A No. 1173	KH-4A	J-15	1678	1964 Nov 2	2130	1013-1	Nov 6	SRV-1 recovered
							1013-2	Nov 7	SRV-2 recovered
89	SS-01A No. 1180	KH-4A	J-16	1705	1964 Nov 18	2035	1014-1	Nov 22	SRV-1 recovered
							1014-2	Nov 27	SRV-2 recovered
90	SS-01A No. 1607	KH-4A	J-17		1964 Dec 19	2110	1015-1	Dec 24	SRV-1 midair recovery
							1015-2	Dec 30	SRV-2 midair recovery
91	SS-01A No. 1608	KH-4A	J-18		1965 Jan 15	2100	1016-1	Jan 20	SRV-1 midair recovery
							1016-2	Jan 25	SRV-2 midair recovery
92	SS-01B No. 1611	KH-4A	J-14		1965 Feb 25	2144	1017-1	Mar 2	SRV-1 midair recovery
							1017-2	Mar 6	SRV-2 midair recovery
93	SS-01B No. 1612	KH-4A	J-19		1965 Mar 25	2111	1018-1	Mar 29	SRV-1 midair recovery
							1018-2	Apr 1	SRV-2 midair recovery
94	SS-01B No. 1614	KH-4A	J-4		1965 Apr 29	2144	1019-1	May 4	SRV-1 midair recovery
							1019-2	May 8	SRV-2 shot to higher orbit
95	SS-01B No. 1615	KH-4A	J-21		1965 May 18	1802	1021-1	May 23	SRV-1 midair recovery
							1021-2	May 28	SRV-2 midair recovery
96	SS-01B No. 1613	KH-4A	J-20		1965 Jun 9	2158	1020-1	Jun 15	SRV-1 midair recovery
							1020-2	Jun 16	SRV-2 recovered from Pacific
97	SS-01B No. 1617	KH-4A	J-22		1965 Jul 19	2201	1022-1	Jul 24	SRV-1 midair recovery
							1022-2	Jul 29	SRV-2 midair recovery
98	SS-01B No. 1618	KH-4A	J-23		1965 Aug 17	2059	1023-1	Aug 22	SRV-1 midair recovery
							1023-2	Aug 26	SRV-2 midair recovery
99	SS-01A No. 1602	R\D	—		1965 Sep 2	2000	—	—	Thor destroyed
100	SS-01B No. 1619	KH-4A	J-24		1965 Sep 22	2131	1024-1	Sep 27	SRV-1 midair recovery
							1024-2	Oct 2	SRV-2 midair recovery
101	SS-01B No. 1616	KH-4A	JX-28		1965 Oct 5	1745	1025-1	Oct 10	SRV-1 midair recovery
							1025-2	Oct 15	SRV-2 midair recovery
102	SS-01B No. 1620	KH-4A	J-25		1965 Oct 28	2117	1026-1	Nov 2	SRV-1 midair recovery
							1026-2	Nov 7	SRV-2 midair recovery
103	SS-01B No. 1621	KH-4A	JX-27		1965 Dec 9	2110	1027-1	Dec 10	SRV-1 midair recovery
							1027-2	Dec 11	SRV-2 midair recovery
104	SS-01B No. 1610	KH-4A	J-26		1965 Dec 24	2106	1028-1	Dec 29	SRV-1 midair recovery
							1028-2	Jan 1	SRV-2 midair recovery
105	SS-01B No. 1623	KH-4A	J-27		1966 Feb 2	2132	1029-1	Feb 7	SRV-1 midair recovery
							1029-2	Feb 12	SRV-2 midair recovery
106	SS-01B No. 1622	KH-4A	J-29		1966 Mar 9	2202	1030-1	Mar 15	SRV-1 midair recovery
							1030-2	Mar 19	SRV-2 midair recovery
107	SS-01B No. 1627	KH-4A	J-30		1966 Apr 7	2202	1031-1	Apr 14	SRV-1 midair recovery
							1031-2	Apr 18	SRV-2 midair recovery

No.[a]	Launch vehicle[h]	Site	Complex[i]		Pad no.	Operation[j]	Orbit[k]	International designation[l]
84	Thor DSV-2C	No. 412	VAFB	75-1	2	2739 KILO KATE	364 × 382 × 115.0	1964-48
85	Thor DSV-2C	No. 405	VAFB	PALC-I	1	3497 QUIT CLAIM	184 × 449 × 85.0	1964-56
86	Thor DSV-2C	No. 421	VAFB	75-3	4	3333 SOLID PACK	184 × 450 × 80.0	1964-61
87	Thor DSV-2C	No. 418	VAFB	PALC-I	1	3559 MOOSE HORN	185 × 436 × 75.0	1964-67
88	Thor DSV-2C	No. 420	VAFB	75-3	4	1013 BROWN MOOSE	184 × 455 × 80.0	1964-71
89	Thor DSV-2C	No. 416	VAFB	75-1	1	3360 VERBAL VENTURE	189 × 369 × 70.0	1964-74
90	Thor DSV-2C	No. 424	VAFB	75-3	4	3358 UTILITY TOOL	181 × 428 × 75.0	1964-85
91	Thor DSV-2C	No. 414	VAFB	75-3	5	3928 BUCKET FACTORY	184 × 441 × 75.0	1965-02
92	Thor DSV-2C	No. 432	VAFB	PALC-I	1	4782 BOAT CAMP	182 × 378 × 75.1	1965-13
93	Thor DSV-2C	No. 429	VAFB	75-3	4	4803 PAPER ROUTE	187 × 278 × 96.0	1965-26
94	Thor DSV-2C	No. 437	VAFB	PALC-I	1	5023 MUSK OX	185 × 482 × 85.0	1965-33
95	Thor DSV-2C	No. 438	VAFB	75-3	4	8431 IVY VINE	203 × 334 × 75.0	1965-37
96	Thor DSV-2C	No. 444	VAFB	75-3	5	8425 FEMALE LOGIC	180 × 368 × 75.1	1965-45
97	Thor DSV-2C	No. 446	VAFB	PALC-I	1	5543 ROCKY RIVER	185 × 472 × 85.1	1965-57
98	Thor DSV-2C	No. 449	VAFB	PALC-I	1	7208 LIGHTS OUT	186 × 427 × 70.0	1965-67
99	Thor DSV-2A	No. 401	VAFB	75-3	5	3373 WORD SCRAMBLE	—	—[m]
100	Thor DSV-2C	No. 458	VAFB	PALC-I	1	7221 NICKLE SILVER	180 × 386 × 80.1	1965-74
101	Thor DSV-2C	No. 433	VAFB	75-3	5	5325 UNION LEADER	208 × 335 × 75.0	1965-79
102	Thor DSV-2C	No. 439	VAFB	PALC-I	1	2155 HIGH JOURNEY	176 × 449 × 75.0	1965-86
103	Thor DSV-2C	No. 448	VAFB	75-3	5	7249 LUCKY FELLOW	183 × 452 × 80.0	1965-102
104	Thor DSV-2C	No. 451	VAFB	75-3	4	4639 TALL STORY	183 × 454 × 80.0	1965-110
105	Thor DSV-2C	No. 450	VAFB	PALC-I	1	7291 SEA LEVEL	184 × 435 × 75.0	1966-07
106	Thor DSV-2C	No. 452	VAFB	75-3	4	3488 EASY CHAIR	181 × 441 × 75.0	1966-18
107	Thor DSV-2C	No. 474	VAFB	PALC-I	1	1612 GAPING WOUND	197 × 322 × 75.1	1966-29

No.[a]	Spacecraft[b]	Pro-gram[c]	Camera	Mass (kg)[d]	Launch date[e]	Time (GMT)	Mission no.[f]	Recovery attempt[g]	Result
108	SS-01B No. 1625	KH-4A	J-28		1966 May 3	1925	1032-1	—	Agena fell in Pacific
							1032-2	—	
109	SS-01B No. 1630	KH-4A	J-33		1966 May 24	0200	1033-1	May 29	SRV-1 midair recovery
							1033-2	Jun 4	SRV-2 midair recovery
110	SS-01B No. 1626	KH-4A	J-31		1966 Jun 21	2131	1034-1	Jun 26	SRV-1 midair recovery
							1034-2	Jul 1	SRV-2 midair recovery
111	SS-01B No. 1631	KH-4A	J-32		1966 Aug 9	2046	1036-1	Aug 16	SRV-1 midair recovery
							1036-2	Aug 23	SRV-2 midair recovery
112	SS-01B No. 1628	KH-4A	J-36		1966 Sep 20	2114	1035-1	Sep 25	SRV-1 midair recovery
							1035-2	Sep 30	SRV-2 midair recovery
113	SS-01B No. 1632	KH-4A	J-38		1966 Nov 8	1953	1037-1	Nov 12	SRV-1 midair recovery
							1037-2	Nov 20	SRV-2 midair recovery
114	SS-01B No. 1629	KH-4A	J-34		1967 Jan 14	2128	1038-1	Jan 19	SRV-1 midair recovery
							1038-2	Jan 26	SRV-2 midair recovery
115	SS-01B No. 1635	KH-4A	J-39		1967 Feb 22	2202	1039-1	Feb 28	SRV-1 midair recovery
							1039-2	Mar 6	SRV-2 midair recovery
116	SS-01B No. 1636	KH-4A	J-35		1967 Mar 30	1854	1040-1	Apr 4	SRV-1 midair recovery
							1040-2	Apr 8	SRV-2 midair recovery
117	SS-01B No. 1634	KH-4A	J-40		1967 May 9	2150	1041-1	May 15	Agena overburn, SRV-1 midair
							1041-2	May 24	SRV-2 midair recovery
118	SS-01B No. 1633	KH-4A	J-37		1967 Jun 16	2135	1042-1	Jun 22	SRV-1 midair recovery
							1042-2	Jul 1	SRV-2 recovered from Pacific
119	Agena D No. 1637	KH-4A	J-42		1967 Aug 7	2144	1043-1	Aug 15	SRV-1 recovered
							1043-2	Aug 22	SRV-2 recovered
120	Agena D No. 1641	KH-4B	CR-1		1967 Sep 15	1941	1101-1	Sep 21	SRV-1 recovered
							1101-2	Sep 28	SRV-2 recovered
121	Agena D No. 1639	KH-4A	J-42		1967 Nov 2	2131	1044-1	Nov 9	SRV-1 recovered
							1044-2	Nov 12	SRV-2 recovered
122	Agena D No. 1642	KH-4B	CR-2		1967 Dec 9	2226	1102-1	Dec 14	SRV-1 recovered
							1102-2	Dec 22	SRV-2 recovered
123	Agena D No. 1640	KH-4A	J-45		1968 Jan 24	2226	1045-1	Jan 31	SRV-1 recovered
							1045-2	Feb 7	SRV-2 recovered
124	Agena D No. 1638	KH-4A	J-48		1968 Mar 14	2200	1046-1	Mar 22	SRV-1 recovered
							1046-2	Mar 29	SRV-2 recovered
125	Agena D No. 1643	KH-4B	CR-3		1968 May 1	2131	1103-1	May 7	SRV-1 recovered
							1103-2	May 15	SRV-2 recovered
126	Agena D No. 1645	KH-4A	J-47		1968 Jun 20	2146	1047-1	Jun 29	SRV-1 recovered
							1047-2	Jul 5	SRV-2 recovered
127	Agena D No. 1644	KH-4B	CR-4		1968 Aug 7	2136	1104-1	Aug 14	SRV-1 recovered
							1104-2	Aug 22	SRV-2 recovered
128	Agena D No. 1647	KH-4A	J-49		1968 Sep 18	2132	1048-1	Sep 27	SRV-1 recovered
							1048-2	Oct 2	SRV-2 recovered
129	Agena D No. 1646	KH-4B	CR-5		1968 Nov 3	2130	1105-1	Nov 12	SRV-1 recovered
							1105-2	Nov 21	SRV-2 recovered
130	Agena D No. 1648	KH-4A	J-50		1968 Dec 12	2222	1049-1	Dec 18	SRV-1 recovered
							1049-2	Dec 23	SRV-2 recovered

No.[a]	Launch vehicle[h]	Site	Complex[i]		Pad no.	Operation[j]	Orbit[k]	International designation[l]
108	Thor DSV-2C	No. 465	VAFB	75-3	5	1508 CARGO NET	—	—
109	Thor DSV-2C	No. 469	VAFB	PALC-I	1	1778 SHORT TON	190 × 293 × 66.1	1966-42
110	Thor DSV-2C	No. 466	VAFB	SLC-1	E	1599 GAME LEG	197 × 375 × 80.1	1966-55
111	Thor DSV-2L	No. 506	VAFB	SLC-1	W	1545 CURLY TOP	193 × 305 × 100.1	1966-72
112	Thor DSV-2C	No. 477	VAFB	SLC-3	W	1703 BIG BADGE	184 × 452 × 85.1	1966-85
113	Thor DSV-2L	No. 507	VAFB	SLC-1	W	1866 BUSY MEETING	175 × 331 × 100.1	1966-102
114	Thor DSV-2C	No. 495	VAFB	SLC-3	W	1664 LONG ROAD	185 × 390 × 80.0	1967-02
115	Thor DSV-2C	No. 493	VAFB	SLC-3	W	4750 BUSY PAWNSHOP	181 × 389 × 80.0	1967-15
116	Thor DSV-2C	No. 501	VAFB	SLC-3	W	4779 GIANT BANANA	167 × 400 × 85.1	1967-29
117	Thor DSV-2L	No. 508	VAFB	SLC-1	E	4696 BUSY BANKER	185 × 799 × 85.1	1967-43
118	Thor DSV-2L	No. 509	VAFB	SLC-1	W	3559	181 × 381 × 80.0	1967-62
119	Thor DSV-2L	No. 510	VAFB	SLC-1	E	4827	174 × 346 × 79.9	1967-76
120	Thor DSV-2L	No. 512	VAFB	SLC-1	W	5089	150 × 389 × 80.1	1967-87
121	Thor DSV-2L	No. 513	VAFB	SLC-1	E	0562	183 × 410 × 81.5	1967-109
122	Thor DSV-2L	No. 514	VAFB	SLC-1	W	1001	158 × 237 × 81.7	1967-122
123	Thor DSV-2L	No. 516	VAFB	SLC-1	E	2243	176 × 430 × 81.5	1968-08
124	Thor DSV-2L	No. 518	VAFB	SLC-1	E	4849	178 × 391 × 83.0	1968-20
125	Thor DSV-2L	No. 511	VAFB	SLC-3	W	1419	164 × 243 × 83.1	1968-39
126	Thor DSV-2L	No. 517	VAFB	SLC-1	E	5343	193 × 326 × 85.0	1968-52
127	Thor DSV-2L	No. 522	VAFB	SLC-3	W	5955	152 × 257 × 82.1	1968-65
128	Thor DSV-2L	No. 524	VAFB	SLC-1	E	0165	167 × 393 × 83.0	1968-78
129	Thor DSV-2L	No. 515	VAFB	SLC-3	W	1315	150 × 288 × 82.2	1968-98
130	Thor DSV-2L	No. 527	VAFB	SLC-3	W	4740	169 × 248 × 81.0	1968-112

No.[a]	Spacecraft[b]	Program[c]	Camera	Mass (kg)[d]	Launch date[e]	Time (GMT)	Mission no.[f]	Recovery attempt[g]	Result
131	Agena D No. 1650	KH-4B	CR-6		1969 Feb 5	2159	1106-1	Feb 9	SRV-1 recovered
							1106-2	Feb 14	SRV-2 recovered
132	Agena D No. 1651	KH-4A	J-43		1969 Mar 19	2138	1050-1	Mar 21	SRV-1 recovered
							1050-2	Mar 22	SRV-2 recovered, no data
133	Agena D No. 1649	KH-4A	J-44		1969 May 2	0147	1051-1	May 9	SRV-1 recovered
							1051-2	May 18	SRV-2 recovered
134	Agena D No. 1652	KH-4B	CR-7		1969 Jul 24	0131	1107-1	Aug 2	SRV-1 recovered from Pacific
							1107-2	Aug 12	SRV-2 recovered
135	Agena D No. 1653	KH-4A	J-46		1969 Sep 22	2111	1052-1	Sep 29	SRV-1 recovered
							1052-2	Oct 7	SRV-2 recovered
136	Agena D No. 1655	KH-4B	CR-9		1969 Dec 4	2137	1108-1	Dec 11	SRV-1 recovered
							1108-2	Dec 21	SRV-2 recovered
137	Agena D No. 1657	KH-4B	CR-10		1970 Mar 4	2215	1109-1	Mar 12	SRV-1 recovered
							1109-2	Mar 23	SRV-2 recovered
138	Agena D No. 1656	KH-4B	CR-11		1970 May 20	2135	1110-1	May 31	SRV-1 recovered
							1110-2	Jun 8	SRV-2 recovered
139	Agena D No. 1654	KH-4B	CR-12		1970 Jul 23	0125	1111-1	Jul 30	SRV-1 recovered
							1111-2	Aug 10	SRV-2 recovered
140	Agena D No. 1658	KH-4B	QR-2R		1970 Nov 18	2129	1112-1	Nov 27	SRV-1 recovered
							1112-2	Dec 7	SRV-2 recovered
141	Agena D No. 1659	KH-4B	CR-13		1971 Feb 17	2004	1113-1	—	Thor destroyed
							1113-2	—	
142	Agena D No. 1660	KH-4B	CR-14		1971 Mar 24	2105	1114-1	Mar 31	SRV-1 recovered
							1114-2	Apr 9	SRV-2 recovered
143	Agena D No. 1662	KH-4B	CR-15		1971 Sep 10	2133	1115-1	Sep 18	SRV-1 recovered
							1115-2	Sep 29	SRV-2 recovered
144	Agena D No. 1661	KH-4B	CR-16		1972 Apr 19	2144	1116-1	May 1	SRV-1 recovered
							1116-2	May 8	SRV-2 recovered
145	SS-01B No. 1663	KH-4B	CR-8R		1972 May 25	1841	1117-1	May 27	SRV-1 recovered
							1117-2	May 31	SRV-2 recovered

No.[a]	Launch vehicle[h]	Site	Complex[i]		Pad no.	Operation[j]	Orbit[k]	International designation[l]
131	Thor DSV-2L	No. 519	VAFB	SLC-3	W	3890	178 × 239 × 81.5	1969-10
132	Thor DSV-2L	No. 541	VAFB	SLC-3	W	3722	179 × 241 × 83.0	1969-26
133	Thor DSV-2L	No. 544	VAFB	SLC-3	W	1101	179 × 326 × 65.0	1969-41
134	Thor DSV-2L	No. 545	VAFB	SLC-3	W	3654	178 × 220 × 75.0	1969-63
135	Thor DSV-2L	No. 531	VAFB	SLC-3	W	3531	178 × 253 × 85.0	1969-79
136	Thor DSV-2L	No. 549	VAFB	SLC-3	W	6617	159 × 251 × 81.5	1969-105
137	Thor DSV-2L	No. 551	VAFB	SLC-3	W	0440	167 × 257 × 88.0	1970-16
138	Thor DSV-2L	No. 555	VAFB	SLC-3	W	4720	162 × 247 × 83.0	1970-40
139	Thor DSV-2L	No. 556	VAFB	SLC-3	W	4324	158 × 398 × 60.0	1970-54
140	Thor DSV-2L	No. 552	VAFB	SLC-3	W	4992	185 × 232 × 83.0	1970-98
141	Thor DSV-2L	No. 537	VAFB	SLC-3	W	3297	—	—
142	Thor DSV-2L	No. 538	VAFB	SLC-3	W	5300	157 × 246 × 81.5	1971-22
143	Thor DSV-2L	No. 567	VAFB	SLC-3	W	5454	156 × 244 × 75.0	1971-76
144	Thor DSV-2L	No. 569	VAFB	SLC-3	W	5640	155 × 277 × 81.5	1972-32
145	Thor DSV-2L	No. 571	VAFB	SLC-3	W	6371	158 × 305 × 96.3	1972-39

Notes to Appendix B

[a]Launch sequence number in CORONA program.

[b]Type and serial number of Agena space vehicle. The satellites were referred to by this number at Lockheed. SS-01A and SS-01B are Standard Stage variants of the Agena D.

[c]Program within CORONA project.

[d]Mass of satellite after orbit insertion.

[e]Dates may differ by one day from sources which give local time.

[f]The CORONA flights were referred to by this classified mission number by the CIA.

[g]Date of SRV recovery, Greenwich Mean Time. Many SRVs were recovered late on the previous date according to local time.

[h]Type and serial number of booster. The DM-18A was similar to the Thor IRBM.

The DM-21 was modified for space use with a shorter guidance section, and the DSV-2A (USAF designation SLV-2) was a standardized version of the DM-21. The DSV-2C (USAF designation SLV-2A) had three Castor strap-on boosters and was also called TAT (Thrust Augmented Thor). Finally, the DSV-2L (USAF designation SLV-2G) had three Castor II solid strapons and a stretched tank; it was also known as LTTAT (Long Tank Thrust Augmented Thor) and Thorad.

[i]VAFB is Vandenberg Air Force Base. NMFPA is Naval Missile Facility, Point Arguello. NMFPA became part of VAFB in 1964. In 1966 the pads were renamed as follows: Complex 75-3 became SLC-1 (Space Launch Complex 1). Complex 75-1 became SLC-2 (Space Launch Complex 2). Complex PALC-1 became SLC-3 (Space Launch Complex 3).

[j]The launches were referred to both by nicknames (until 1967) and by OPS number. After 1963, the OPS number was also used to refer to the orbiting satellite in open sources; thus CORONA flight 145 appears in the NORAD satellite catalog as OPS 6371.

[k]The orbit is given with perigee and apogee in kilometers followed by inclination to the eastward equator in degrees. The orbital heights for flights 1 to 118 are taken from the Vandenberg launch summary reports and are slightly different from the values given in the Royal Aircraft Establishment tables which are usually quoted.

[l]This is the international designation assigned by the World Data Center, Rockets and Satellites for COSPAR.

[m]*Notes on individual launches:*

1. Flight 1 may not have reached orbit. It did not carry an SRV.

2. Flights 19 and 21 were infrared sensor tests and did not carry SRVs. Flight 21 demonstrated on-orbit restart of Agena, raising apogee by 300 km.

3. Flight 37 carried the last KH-3 camera in an engineering model of the KH-4/Mural spacecraft; the launch photograph clearly shows the distinctive KH-4 spacecraft bus.

4. Flights 54 and 99 carried scientific payloads with no SRV.

5. Flight 64 suffered a payload power failure and took no pictures.

6. Flight 78 Agena reentered over Venezuela with both SRVs still attached on May 26; SRV-1 survived reentry.

USAF Program Numbers:

Flights 1 to 39 were under the DISCOVERER program. Flights 40 to 48 were under Program 622A.

Flights 49 to 78 were under Program 162. Flights 79 to 112 were under Program 241. Flights 113 to at least 118 were under Program 846.

Note: The author would be pleased to hear from any reader who can provide corrections or fill in the missing spacecraft masses. Contact Jonathan McDowell, Harvard–Smithsonian Center for Astrophysics, Cambridge, MA 02138.

NOTES

INTRODUCTION

1. Johnson's talk was not recorded and there are differing accounts of exactly what he said. This account, except for the last sentence, comes from William Burrows, *Deep Black* (New York: Random House, 1986), viii. Evert Clark's *New York Times* article of the event does not contain this quote, however ("Satellite Spying Cited by Johnson," March 17, 1967, 13). The last sentence of the quote is from an account by the National Photographic Interpretation Center contained in a briefing chart and listed as "The Nashville Address."
2. National Science and Technology Council, The White House, "Fact Sheet: National Space Policy," September 19, 1996, 7.
3. The term *photoreconnaissance* as used in this volume means obtaining imagery in the visual range from earth orbit. In the early years of the U.S. photoreconnaissance program, this imagery was recorded on film in the orbiting satellite, and then returned to Earth. Since at least the mid-1970s, as acknowledged by a September 19, 1996, National Space Policy statement, imagery is returned to Earth electronically in "near real-time." The space policy statement also acknowledged that the United States, in addition to conducting photoreconnaissance from orbit, carries out "overhead signals intelligence collection" and "overhead measurement and signature intelligence collection." Ibid., 7–8. This volume discusses only the origins and evolution of U.S. photoreconnaissance activities.
4. Another version of this story is that the name came from the Corona brand of cigars; however, the most frequently cited version is the typewriter story.
5. All of this film is now available for viewing at the National Archives and Record Administration (NARA). See Appendix B for a complete listing of CORONA missions.
6. By comparison, the total land area of the Soviet Union was 8.6 million square miles and the total land area of the earth is 58 million square miles.

7. Robert Louis Benson and Michael Warner, eds., *VENONA: Soviet Espionage and the American Response, 1939–1957* (Washington, D.C.: National Security Agency and Central Intelligence Agency, 1997).
8. Kevin Ruffner, ed., *CORONA: America's First Satellite Program,* Center for the Study of Intelligence (Washington, D.C.: Government Printing Office, 1995), xiii.
9. Kevin Ruffner, "CORONA and the Intelligence Community," *Studies in Intelligence* 39, no. 5 (1996): 62.
10. Ibid.
11. "Executive Order 12951 of February 22, 1995: Release of Imagery Acquired by Space-based National Intelligence Reconnaissance System," *Federal Register* 60, no. 39 (February 28, 1995): 10789–90.
12. Kenneth Greer, "CORONA," as reprinted in *CORONA,* ed. Ruffner.
13. See chapter 2, "CORONA: A Triumph of American Technology," by Albert Wheelon.

CHAPTER 1. STRATEGIC INTELLIGENCE AND U.S. SECURITY

1. John Lewis Gaddis, *Strategies of Containment: A Critical Appraisal of Postwar American National Security Policy* (New York: Oxford University Press, 1982), remains the best survey of American strategic debates in the period, but is partially superseded by Robert R. Bowie and Richard H. Immerman, *Waging Peace: The National Security Policies of the Eisenhower Administration* (forthcoming).
2. "NIE 11-6-54: Soviet Capabilities and Probable Programs in the Guided Missile Field," in *Estimates on Soviet Military Power, 1954 to 1984,* ed. Donald Steury (Washington, D.C.: Center for the Study of Intelligence, 1996).
3. "NIE 11-8-60: Soviet Capabilities and Probable Programs in the Guided Missile Field," in *Intentions and Capabilities: Estimates on Soviet Strategic Forces, 1950–1983,* ed. Donald P. Steury (Washington, D.C.: Government Printing Office, 1996).
4. "NIE 11-8/1-61: Strength and Deployment of Soviet Long-Range Ballistic Missile Forces," in *Intentions and Capabilities,* ed. Steury (Air Force Intelligence dissented, insisting that Soviet ICBMs were and would be more numerous, but it was alone in this view).
5. "NIE 11-8-62: Soviet Capabilities for Long-Range Attack," in *Intentions and Capabilities,* ed. Steury.
6. "NIE 11-8-64: Soviet Capabilities for Strategic Attack," in *Intentions and Capabilities,* ed. Steury.
7. Kevin Ruffner, ed., *CORONA: America's First Satellite Program,* Center for the Study of Intelligence (Washington, D.C.: Government Printing Office, 1995), 24.
8. "Photographic Interpretation Report: KH-4 Mission 1042-1, 17–22 June 1967," in *CORONA,* ed. Ruffner, 261–88.
9. See U.S. Department of State, *Foreign Relations of the United States, 1961–1963,* vol. 14: *Berlin Crisis, 1961–1962* (Washington, D.C.: Government Printing Office, 1993), passim.

10. This is a point developed in a book by Timothy Naftali and Alexander Fursenko analyzing both the American and Soviet sides of the Berlin and Cuban missile crises. See *"One Hell of a Gamble": Khruschev, Castro, and Kennedy, 1958–1964* (New York: Norton, 1997).
11. National Security Council Executive Committee meeting, October 25, 1962, John F. Kennedy Library, President's Office Files, Presidential Recordings Collection: Meeting Recording No. 37.4 (declassified October 1996).
12. The best account of U.S. force planning and arms control negotiation is McGeorge Bundy, *Danger and Survival: Choices about the Bomb in the First Fifty Years* (New York: Random House, 1988).

CHAPTER 2. CORONA

Note: Based on the keynote address by the author presented on May 23, 1995, at "Piercing the Curtain: CORONA and the Revolution in Intelligence," conference sponsored by the George Washington University and the Central Intelligence Agency.

1. Edwin Land, personal communication to Albert D. Wheelon, n.d.
2. J. E. Lipp and R. M. Salter, "Project Feed Back Summary Report," RAND Corporation, Contract No. AF 33(038)-6413 (March 1, 1954).
3. RCA was responsible for developing the television sensor system but recommended that it be abandoned in August 1957.
4. It is significant that it was Land who informed Bissell that he had been given responsibility for the CORONA project, thereby indicating the extraordinary influence that these advisors wielded.
5. The SAMOS system that many hoped would replace CORONA was canceled a few years later.
6. Two SR-71s reentered Air Force service in the mid-1990s.
7. Then known as Camp Cooke and now called Vandenberg Air Force Base. The 672nd Strategic Missile Squadron (Thor) was based there. Range instrumentation and tracking facilities were also available.
8. At the time, it was referred to as the Hustler.
9. Their HYAC camera was developed for the Air Force WS-461L Balloon Reconnaissance Program and is now on display in the Smithsonian National Air and Space Museum. It was a panoramic camera with a 12-inch focal length.
10. This was more difficult than the previous plan of using a spinning spacecraft to provide attitude control through gyroscopic stability, although the camera was simpler and alaready proven.
11. 175 line pairs per millimeter at 2-to-1 contrast.
12. I. G. Kolchinskii, "Optical Instability of the Earth's Atmosphere According to Stellar Observations," originally published in Russian in *Naukova Dumka* (Kiev, 1967); translated into English by the Aeronautical Chart and Information Center, St. Louis, Missouri (February 1969).
13. R. E. Hufnagel and N. R. Stanley, "Modulation Transfer Function Associated with Image Transmission Through Turbulent Media," *Journal of the Optical Society*

of America 54, no. 1 (January 1964): 52–61; V. I. Tatarskii, "The Effect of the Turbulent Atmosphere on Wave Propagation," translated from Russian and issued by the National Technical Information Office, U.S. Department of Commerce (1971).

14. In the early years of CORONA, there was great concern about the reserve of control gas available in the Agena, and also concern about potential damage to the film-return capsule from impact with particles in space. To minimize these risks, the Agena performed a 180-degree yaw maneuver immediately after achieving orbit. Thus, a 60-degree tilt down was the maneuver required to de-orbit. As more was learned about space and the reliability of the Agena improved, the yaw maneuver was eliminated and the recovery tilt down became 120 degrees. This allowed the main Agena engine to be restarted, if desired, to accomplish a change of orbit during the flight.
15. The rocket was fired after tilting down 60 degrees from the forward direction to minimize the influence of alignment errors on the impact zone.
16. This daring concept had been attempted in recovering reconnaissance balloons, usually with discouraging results.
17. Ironically, this first film capsule was returned on August 18, 1960, the same day that Gary Powers was sentenced in Moscow for the last U-2 overflight of the USSR.
18. The breaking of the German cipher machine codes produced communications intelligence of extraordinary importance and was also known as ULTRA.
19. The first U.S. thermonuclear device was exploded during the Mike test on November 1, 1952, at Eniwetok Atoll in the South Pacific. This weapon weighed 60 metric tons, and could not be carried by an airplane. On August 12, 1953, the Soviets tested a device with a yield of 400 kilotons, about 25 times smaller than the Mike test. While technically not a hydrogen bomb, this boosted-fission weapon could be made into an operational bomb far more easily than the American version. On November 20, 1955, the Soviets tested their first true thermonuclear bomb, which was dropped from a Tu-16 bomber. The test had a yield of 1.6 megatons, but the bomb easily could have been modified to yield twice that. David Holloway, *Stalin and the Bomb* (New Haven: Yale University Press, 1994), 299–315.
20. When 1961 arrived, we found only six Soviet ICBMs deployed. Kennedy realized that we were well ahead. The United States had frightened itself through ignorance.
21. "Piercing the Curtain."
22. William E. Burrows, *Deep Black: Space Espionage and National Security* (New York: Random House, 1986), 207–9.
23. Air Force Major Rudolf Anderson became the only casualty of this crisis. Anderson was shot down by a Soviet-operated SA-2 in a CIA U-2, just as Gary Powers had been downed two years before.
24. There is some dispute about who solved this problem. Frank Madden of Itek believes that Itek's materials department was primarily responsible for the solution. See chapter 3 for further details.

25. Albert D. Wheelon, "Antisatellite Weapons and Space Warfare," *Annals of the New York Academy of Sciences* 489 (December 26, 1986): 38–47, from the symposium, "The High Technologies and Reducing the Risk of War," held on May 7–8, 1986.
26. The Soviet ASAT succeeded in eleven of twenty-two tests against orbital targets.

CHAPTER 3. THE DEVELOPMENT AND IMPROVEMENT OF THE CORONA SATELLITE

Note: The author would like to acknowledge the assistance of the following people who reviewed copies of the manuscript at various points during its preparation: Jim Plummer, Robert Hopkins, Bob Leeper, Walter Levison, Bill King, Charlie Murphy, and Frank Buzard. He is particularly indebted to Frank Madden, whose history of the camera development at Itek was invaluable.

1. Richard Bissell, in "A Point in Time," CIA-produced CORONA documentary filmed in 1972 at the end of the program. This documentary video was declassified in 1995 and is available to the public through the National Technical Information Service.
2. Kenneth E. Greer, "CORONA," *Studies in Intelligence,* Suppl. 17 (Spring 1973): 1–37, in *CORONA: America's First Satellite Program,* ed. Kevin C. Ruffner, Center for the Study of Intelligence (Washington, D.C.: Government Printing Office, 1995), 8.
3. Greer, "CORONA," 9–10.
4. James Plummer, comments at "Piercing the Curtain," a joint symposium held by the CIA's Center for the Study of Intelligence and the George Washington University's Space Policy Institute at the George Washington University, Washington, D.C., on May 23–24, 1995.
5. Frank Buzard, letter to Dwayne Day, November 1996.
6. WS-117L also included another program that was used for infrared early warning of ballistic missile attack and was eventually named MIDAS. Two of the Discoverer launches were then allocated to serve as radiometric vehicles which would ostensibly be used for space navigation purposes, but whose data would actually be used for the development of MIDAS.
7. The launch windows generally opened around noon for the early missions. As the spacecraft's lifetime increased, the launch window depended more upon the area of primary interest, which had to be well lit when the satellite overflew it.

   Charlie Murphy, who was eventually responsible for satellite launch operations at Vandenberg, states that the railroad schedule was not as big a problem as people thought. According to Murphy, one of the first things the new launch officer did was to go to railroad headquarters in San Francisco and try to convince the railroad scheduler to be more flexible. The scheduler would usually pat the young man on the head and ignore him. Murphy does not recall a single mission that had to slip to another day because of a train passing through the base.
8. The Agena was originally named Hustler, after the Bell Hustler rocket engine that Lockheed had selected to power it. The rocket itself was named after the B-58 Hustler aircraft. It was supposed to power a weapons pod underneath the aircraft,

but the pod was eventually canceled. The upper stage was renamed Agena sometime in 1958 (there are no records of this decision), apparently by an ARPA committee.

9. The spacecraft had to provide a temperature range of plus or minus 10 degrees Fahrenheit from the optimum design temperature of 70 degrees Fahrenheit. Frank J. Madden, "The CORONA Camera System: Itek's Contribution to World Security," self-published monograph, October 1996, 10, 20.
10. Buzard, letter to Day, November 1996.
11. Greer, "CORONA," 13.
12. Bob Leeper stated that this signal was very distinctive, sort of a high-frequency repetitive sweeping sound that could be picked out of the background noise. Bob Leeper, letter to Dwayne Day, November 23, 1996.
13. There was at least one incident where this did not work as planned. During an early CORONA mission the recovery vehicle came down off-target, landing in warm waters near Christmas Island. The plug dissolved faster than planned and the bucket sank before a recovery crew could retrieve it.
14. Each of the launches was given an operational name. Examples were: FLYING YANKEE, FROGGY BOTTOM, CHILI WILLIE, and GIANT BANANA. (See Appendix B.)
15. Greer, "CORONA," 14.
16. Frank Buzard, interview by Dwayne A. Day, June 21, 1996.
17. "A Point in Time."
18. Greer, "CORONA," 16.
19. N. F. Twining, Chairman, Joint Chiefs of Staff, Memorandum for the Deputy Secretary of Defense, "Discoverer Capsule," April 25, 1959, NARA, JCS Files.
20. A. Roy Burks, interview by Dwayne A. Day, May 1996.
21. For instance, the thickness of the film determined how much could be packed into the bucket. More important, the lifting capability of the Thor-Agena combination established a rigid limit on how much film could be carried regardless of the size of the film-return bucket. In addition, the bucket would have provided no evidence of the width of the film; early versions of CORONA carried only one film spool, whereas the KH-4 and later spacecraft carried two—one for each camera.
22. This episode, as two superpowers frantically searched for a photoreconnaissance satellite vehicle lost in the Arctic, later served as the basis for the book and movie *Ice Station Zebra*.

    As for the Soviet fishing trawler? "That same trawler could have 'suddenly left port' on schedule every Saturday afternoon." Buzard interview.
23. Greer, "CORONA," 17.
24. Ibid.
25. Merton E. Davies and William R. Harris, *RAND's Role in the Evolution of Balloon and Satellite Observation Systems and Related U.S. Space Technology,* R-3692-RC (Santa Monica, Calif.: RAND Corporation, 1988), 81.
26. Robert A. McDonald, "CORONA: Success for Space Reconnaissance, A Look into the Cold War, and a Revolution for Intelligence," *PE&RS* 61 (June 1995): 694–95.

27. There was a horizon photo on each end, so about 33 inches of film was transported during each cycle.
28. Madden, "CORONA Camera System," 13.
29. Later, Itek hired and trained enough craftsmen to fabricate all elements internally. The production rate increased from one lens every two months to eight lenses every month (ibid., 19–22).
30. McDonald, "CORONA," 718.
31. In 1954, Kodak film expert Henry Yutzy told Edwin Land about a new thin-based Mylar film under development at Kodak. Yutzy said that Kodak saw little future for such films, but Land knew that such films were vital to aerial reconnaissance and convinced Yutzy that Kodak should continue such research out of patriotic duty. Donald E. Welzenbach, "Science and Technology: Origins of a Directorate," *Studies in Intelligence* 30 (Summer 1986): 25, found in RG 263, NARA.
32. The performance of a rocket as it exhausts its fuel becomes somewhat unpredictable. The Atlas rocket had sufficient power to spare and could be shut down before running out of fuel, thus allowing a more accurate orbit to be attained. The Thor, however, was already being pushed to its limits early in the CORONA program.
33. Greer, "CORONA," 19.
34. Col. C. L. Battle, USAF, and J. W. Plummer, "The Discoverer Program," Lockheed Missiles and Space Company, 2175-61, [n.d.], 13.
35. Greer, "CORONA," 20–21.
36. "CORONA Program Profile," Lockheed Press Release, May 1995. It should also be noted that the morale of those working on the program is somewhat disputed by those were involved at the time. Several people, such as Charlie Murphy and Frank Buzard, have stated that they did not think the program was in jeopardy at that time. Clearly, however, CORONA was using up its initial batch of allocated Thor booster rockets and the SAMOS program was already conducting launches, albeit without success.
37. Greer, "CORONA," 21.
38. Buzard interview.
39. Ibid.
40. According to Charlie Murphy, Jim Plummer was responsible for installing the flag aboard the capsule. Plummer did this under a classified Air Force contract—something that he found quite amusing.
41. The CORONA development team in Palo Alto celebrated at a local hotel and threw Jim Plummer into the swimming pool.
42. Usually the entire bucket was transferred to the West Coast sealed inside a light-tight bag and a large steel drum. Light-sensitive film was placed inside the bag to help determine if any accidental exposure occurred during the trip.
43. As the missions became routine, the film canisters were shipped east with a CIA courier via commercial airline flights. Bob Leeper, interview by Dwayne A. Day, June 27, 1996.
44. "A Point in Time," and Greer, "CORONA," 24.
45. Bill Crimmins, interview by Dwayne A. Day, August 8, 1996.

46. Clouds were always a worry for satellite photoreconnaissance. Early CORONA missions averaged 50–60 percent cloud cover. This improved over the lifetime of the program due to two reasons: improved satellite control technology, and better weather forecasting equipment and models. For early CORONA missions, there was no way to alter the camera switch-on and -off commands after launch. The commands were all on a piece of Mylar tape. If controllers knew that there was going to be heavy cloud cover over a target, they could not do anything about it. This was even more problematic because during the early days of the program they could not accurately control the period of the satellite (the time it took to orbit the earth, which was determined by the altitude). The period determined the satellite's ground track. The result was that a satellite could be a considerable distance from where it was supposed to be after launch and the controllers would not know precisely what territory it was covering at any specific moment. Eventually the Lockheed team, led by Jim Plummer, was able to insert a command function into the spacecraft that allowed controllers to tell it to obey or ignore a camera-on command on the Mylar tape. As the launch vehicle became more powerful, the satellite's controllers were able to achieve more precise orbital periods as well, and the Agena B upper stage included small motors that allowed them to place the satellite in a more precise orbit. Finally, ground-based satellite tracking improved. All of these improvements dramatically increased the productivity of the missions. If controllers knew there were clouds over the target and knew that the satellite was going to pass over that area, they simply ordered the spacecraft to ignore the scheduled command to turn on the cameras. This was the usual practice, but if the target had a high priority, they took the photos anyway and hoped for the best. By the end of the CORONA program, cloud cover had been reduced to around 27 percent. Dino Brugioni, interview by Dwayne A. Day, November 6, 1996, and Charlie Murphy, interview by Dwayne A. Day, December 5, 1996.
47. A popular story is that word apparently came down from Washington that all information about the flight was to be suppressed. Those in possession of the capsule took this to mean that the vehicle should be destroyed and so they altered the bucket with sledgehammers and then dropped the battered capsule in San Francisco Bay. A different, apparently phony bucket was later presented to the U.S. Air Force Museum at Wright-Patterson Air Force Base in Dayton, Ohio, where even today it is still labeled "Discoverer XIV." According to several of those involved in the program, the story is simply not true and the capsule in the Air Force museum is indeed that from the successful film mission. Some components of the buckets, such as the radio beacons, were eventually re-used, but the buckets were cut up and dumped in the ocean, according to several sources.
48. *CORONA,* ed. Ruffner, 41–42.
49. Allen W. Dulles, Director of Central Intelligence, Director of Central Intelligence Directive No. (Deleted), Committee on Overhead Reconnaissance (COMOR), August 9, 1960, contained in *CORONA,* ed. Ruffner, 43.

50. The "Mushroom Factory" earned its nickname from those who worked there for two reasons: it was always damp in the Pentagon basement where their offices were located, and there was never any shortage of "fertilizer" from the upper levels of the bureaucracy. Charlie Murphy, interview by Dwayne A. Day, March 7, 1996.
51. *CORONA,* ed. Ruffner, 97.
52. James Q. Reber, Chairman, Committee on Overhead Reconnaissance, "List of Highest Priority Targets: USSR," August 18, 1960, contained in *CORONA,* ed. Ruffner, 49–58.
53. Col. James E. Mahon, Memorandum for COMOR, "Urgent Requirements for CORONA and ARGON," August 18, 1960, contained in *CORONA,* ed. Ruffner, 45–46. ARGON was sponsored by the Army Mapping Service and its camera was manufactured by Fairchild. ARGON carried so much film that during its operation the procedure was to simply turn the camera on and leave it on for the entire flight.
54. McDonald, "CORONA," 694.
55. Bill King, interview by Dwayne A. Day, March 28, 1996.
56. McDonald, "CORONA," 719, 694. In actuality, the problem was the difficulty of attaining the predicted orbit. The new camera could be adjusted to match the actual orbit.
57. Col. C. L. Battle, USAF, and J. W. Plummer, "The Discoverer Program," Lockheed Missiles and Space Company, 2175-61, [n.d.], 6.
58. The Agena already included ullage rockets to start its main engine just after separation from the Thor during launch.
59. Madden, "CORONA Camera System," 16.
60. Walter Levison, comments at "Piercing the Curtain," May 1995. Information on the design of the camera comes from Frank Madden, in a letter to Dwayne Day, November 1996.
61. Madden, "CORONA Camera System," 19.
62. McDonald, "CORONA," 694–95, 718, and Greer, "CORONA," 28.
63. Two other CORONA missions also did not carry cameras. One was the STARAD radiation detection mission launched in October 1962 and the other was an R&D mission launched in September 1965 which failed to reach orbit.
64. Stereo viewing is not the only way to obtain height data. It is also possible to measure an object's shadow and determine its height using simple mathematics. Data could also be obtained using the known distance to the target as well as other ephemera, like the sun angle. But these alternative methods relied upon an assumption that the ground around the target was perfectly flat. Even slight changes in the ground slope could produce dramatically inaccurate measurements. Stereo-imagery was far superior. See Dino Brugioni's comments in chapter 11 for further comments on this.
65. Madden, "CORONA Camera System," 34. MURAL also allowed the elimination of the momentum wheel, which saved weight.
66. Greer, "CORONA," 29–30.
67. There is some question about whether this failure carried the last C‴ camera or the first MURAL. According to several sources, this mission, number 9030,

carried a single C‴ camera. See, for instance, "Satellite Photographic Reconnaissance Flight Summary," CORONA Briefing Book. But Jonathan McDowell, who keeps extensive lists of world satellite launches (and who provided the launch listing in this book), located a launch photograph of this mission which shows a vehicle with a much larger payload than that carried for C‴ missions, but was the same size as that used for the MURAL missions. The tail number of the Thor rocket matches the correct mission, leaving open only two possibilities: Mission 9030 actually carried a MURAL camera instead of a C‴, or it carried a single C‴ camera in a larger payload fairing. Because this mission was a failure, however, it did not change the numbers of either KH-3 or KH-4 satellites actually placed in orbit.

68. The secrecy was also extended to all other U.S. military space launches, including those of very innocuous and well-publicized programs like the Transit navigation satellites. This secrecy proved to be troublesome and expensive for many of the nonreconnaissance programs.
69. Greer, "CORONA," 31. Attaining the optimum atitude was a major concern for satellite controllers and engineers alike for the early satellites. Low orbits increased resolution but decreased the amount of territory covered. High orbits did the opposite. Photo-interpreters generally preferred higher resolution, but in the early days they were happy just to be seeing previously uncovered areas. Charlie Murphy, the Air Force officer detailed to the CIA who was in charge of operations at AP, usually chose to go for a higher altitude for the early satellites. The Lockheed engineers appreciated this because it was difficult to determine the altitude of the satellites to begin with and higher altitudes decreased drag on the satellite and decreased the chances that it would come down in the wrong place.
70. The information on the Index camera, along with the story about the whale oil, comes from Madden, "CORONA Camera System," 45–46.
71. Ibid., 53.
72. Exactly who was responsible for the fix is a matter of some dispute. Albert Wheelon, then deputy director of science and technology at the CIA, remembers asking Sydney Drell of Stanford to form a group to evaluate this phenomenon. The Drell group, according to Wheelon, then went on to address other technical issues concerning CORONA. Frank Madden of Itek says that the problem was solved by the Itek environmental test lab people and that Dr. Edward Purcell, a member of the Drell group, visited the lab to check on their work. Madden wrote, "After his examination, he asked if our solution seemed to be working and we said it appeared to be. His rejoinder was that we had best stick with it. He noted that we basically had a Van de Graaff generator, and that he had never been able to understand them. We all gave him great credit for accepting an experimental solution as valid, even if it was not able to be substantiated theoretically": Madden, "CORONA Camera System," 16. Bob Leeper remembers the Drell group as a Blue Ribbon Committee formed to analyze the overall CORONA system from a product quality standpoint, not specifically to address the electrostatic discharge problem. Leeper, letter to Day.
73. Frank Madden, interview by Dwayne A. Day, November 17, 1995.

74. These precautions included starting the film one or two frames earlier during each photo pass. Ibid.
75. Brugioni interview.
76. Carl Berger, *The Air Force in Space, Fiscal Year 1961* (Washington, D.C.: USAF Historical Division Liaison Office, 1966), 34–38.
77. Dudley C. Sharp, Secretary of the Air Force, Memorandum for the Chief of Staff, USAF, September 13, 1960, Thomas D. White Papers, Box 36, "4–5 Missiles/Space/Nuclear," Library of Congress (hereinafter cited as LC).
78. "SAMOS Special Satellite Reconnaissance System," USAF, n.d., Office of the Staff Secretary: Records of Paul T. Carroll, Andrew J. Goodpaster, L. Arthur Minnich, and Christopher H. Russell, 1952–61, Subject Series, Alphabetical Subseries, Box 15, "Intelligence Matters (14) [March–May 1960]," Dwight D. Eisenhower Library. Also, "CORONA Program Profile."
79. Brig. Gen. John M. Breit, USAF, Deputy Inspector General for Security, to AFCCS, "Unauthorized Disclosure of Classified Information (The SAMOS Reconnaissance Satellite)," October 23, 1960, Thomas D. White Papers, Box 36, "4–5 Missiles/Space/Nuclear," LC. SAMOS E-4 was a mapping camera canceled when the decision was made to pursue ARGON. SAMOS E-3 was apparently a 36-inch camera canceled early on as well. Buzard, letter to Day.
80. Although the electronic intelligence aspects of the U.S. military space program remain classified, there have been a number of declassified documents concerning this SAMOS mission. See, for instance, Hugh S. Cumming Jr. to Mr. Farley, "SAMOS," July 22, 1960, General Records of the Department of State, Bureau of European Affairs, Office of Soviet Union Affairs, Subject Files, 1957–63, Box 6, "12 Satellite and Missile Programs," RG 59, NARA. See also Lt. General Roscoe C. Wilson, Deputy Chief of Staff, Development, USAF, to Air Research and Development Command, "Exploitation of Initial SAMOS Data," June 1, 1960, Thomas D. White Papers, Box 34, LC.
81. Lt. Gen. Bernard A. Schriever, Air Research and Development Command, to Gen. Thomas D. White, Chief of Staff, USAF, September 13, 1959, Thomas D. White Papers, Box 26, LC.
82. Lt. Gen. Bernard A. Schriever, Air Research and Development Command, to Gen. Thomas D. White, Chief of Staff, USAF, August 1, 1959, Thomas D. White Papers, Box 26, LC.
83. Gen. Curtis LeMay, Deputy Chief of Staff, USAF, to Lt. Gen. Bernard Schriever, Air Research and Development Command, September 9, 1959. Thomas D. White Papers, Box 26, LC.
84. Buzard interview.
85. Lee Bowen, *The Threshold of Space: The Air Force in the National Space Program, 1945–1959* (Washington, D.C.: USAF Historical Division Liaison Office, September 1960), 30–32, 33.
86. King interview.
87. Harold F. "Bud" Wienberg, interview by R. Cargill Hall, March 16, 1995.
88. Welzenbach, "Science and Technology," 24.
89. Leeper interview.

90. Greer, "CORONA," 31. The "SAMOS Special Satellite Reconnaissance System" report identifies the E-5 as having a 66-inch focal length, which is identical with the KH-6 LANYARD.
91. McDonald, "CORONA," 716; Greer, "CORONA," 31.
92. The KH-4A was sometimes referred to as the KH-4J in reference to its J-1 camera.
93. Madden, "CORONA Camera System," 36–38.
94. Greer, "CORONA," 32.
95. Madden, "CORONA Camera System," 34.
96. McDonald, "CORONA," 695. CORONA missions were usually launched around noon to mid-afternoon. This allowed them to photograph the Soviet Union on their descending passes. In mid-summer, so much of the Soviet Union was in sunlight for most of the time that the satellite could photograph targets both on its descending (north-south) and ascending (south-north) passes. The ascending passes were usually at a higher altitude, since the satellite's orbit was at its highest over the South Pole and thus the satellite was traveling south to north while also decreasing in altitude. While this lowered the resolution for these ascending passes, decent imagery was useful no matter what and controllers took this opportunity whenever they had it.
97. Greer, "CORONA," 32–33.
98. Burks interview.
99. "Declassified Imaging Satellite Systems," CIA Fact Sheet.
100. Organization Chart, Office of Special Activities—Deputy Director (Research), attached to Col. Stanley W. Beerli, USAF, Assistant Director for Special Activities, Office of Special Activities, OSA HQS Notice No. 1–15, June 29, 1962, RG 263, NARA.
101. Buzard interview. The CIA also had a security officer at AP. The details of this time period are still somewhat in dispute.
102. Welzenbach, "Science and Technology," 24.
103. For information on the TFX effort, see Robert J. Art, *The TFX Decision: McNamara and the Military* (Boston: Little, Brown, 1971). For information on the Gemini controversy, see Barton C. Hacker and James M. Grimwood, *On the Shoulders of Titans: A History of Project Gemini* (Washington, D.C.: Government Printing Office, 1977), 118–22.
104. Whether McNamara did this is unclear. However, it is consistent with other actions by the secretary of defense at the time. According to one account of NASA's history during this same time period, McNamara proposed that DoD take over all manned spaceflight in earth orbit, leaving NASA to conduct missions beyond earth orbit. See Arnold S. Levine, *Managing NASA in the Apollo Era* (Washington, D.C.: Government Printing Office, 1982), 230.
105. Albert "Bud" Wheelon, interview by Dwayne A. Day, June 22, 1996.
106. Welzenbach, "Science and Technology," 26.
107. Albert Wheelon, comments at "Piercing the Curtain," May 23, 1995.
108. Wheelon states: "The debate between CIA and DoD then shifted to whether CIA ought to pursue new reconnaissance systems. Assistant Secretary of Defense

Gene Fubini and Brockway MacMillan argued against each system that CIA was developing. The debate continued until Al Flax became director of the NRO in 1965. He saw CIA and the Air Force as valuable and complementary assets. This was fortunate for the country because the three systems then being developed at CIA would eventually become vital components of the National Reconnaissance Program." Albert D. Wheelon, "Lifting the Veil on CORONA," *Space Policy* (November 1995): 253.

109. According to Charlie Murphy, the issue was partly one of the Air Force not understanding why the CIA was even involved in satellites. The CIA was supposed to be about *spies,* not about expensive pieces of equipment. During the early years of CORONA most of the design and engineering work for the payload was conducted by contractors. Later, the CIA began to acquire its own in-house engineering capability (see chapter 6) and this mystified many in the Air Force, who had long experience with designing and procuring large technological systems. Murphy interview, December 5, 1996.
110. Buzard interview.
111. The person the CIA sent out was A. Roy Burks, who is seen in a number of the photographs throughout this book. Burks interview.
112. Lockheed apparently was offered money by both the CIA and the Air Force, but refused to accept it until the dispute was formally resolved. Murphy interview, December 5, 1996.
113. Leeper interview.
114. Murphy interview, December 5, 1996.
115. Greer, "CORONA," 34.
116. With the KH-4B, Lockheed allowed an extra two pounds for the lens. This made lens production significantly easier since the lens elements could be made a little thicker and were less prone to bending during polishing and test. Element fabrication time was significantly reduced. Madden, "CORONA Camera System," 20.
117. Greer, "CORONA," 36. Frank Madden indicates that some ultrathin-base film was spliced to the end of a roll on an operational mission and flown, but the imagery was very variable. Transport problems emerged in the lab and negated its use. Madden letter.
118. Madden letter.
119. The resolution figures cited for the CORONA cameras represent the best figures obtained from orbit. This was usually determined from examining a resolution test target on the ground in Arizona, where the thin, clear air reduced distortion. In actual practice, resolution could vary considerably depending on a wide number of factors. The initial limitations were caused by imprecise orbits (especially when the cameras had fixed image motion compensation) and camera vibration. Later the problems were mostly the result of atmospheric conditions, such as pollution, haze, or different air temperatures (and thus density) at different altitudes.
120. Because of the extreme lighting conditions encountered during a mission (film could be overexposed under conditions of snow, ice, desert or near the equator and underexposed at far northern latitudes or low sun angles), the film had to be developed to different levels. In order to determine the amount of development

needed, Eastman Kodak came up with an infrared inspection method that could be used to determine optimum exposure levels so that additional development could be applied before the image was fixed.

121. McDonald, "CORONA," 707–8, and Report No. 9, KH-4B System Capability, "Appraisal of Geologic Value for Mineral Resources Exploration," March 1971, contained in *CORONA,* ed. Ruffner, 321.
122. Greer, "CORONA," 36–37.
123. "Appraisal of Geologic Value for Mineral Resources Exploration," 321.
124. *CORONA,* ed. Ruffner, 356.
125. Linda Neuman Ezell, *NASA Historical Data Book,* vol. 3: *Programs and Projects 1969–1978* (Washington, D.C.: NASA, 1988), 335–41.
126. Walter Levison, interview by Dwayne A. Day, November 16, 1995.
127. Albert Wheelon, comments at "Piercing the Curtain," May 23, 1995.
128. Murphy interviews, March 7 and December 5, 1996.
129. Greer, "CORONA," 39.
130. According to Frank Madden, the qualification models for the KH-3, KH-4, and KH-4A were all refurbished and flown. For the KH-4B qualification model, the government funded only one camera of the stereo pair and the central supporting structure. This is the model on display in the National Air and Space Museum, Washington, D.C. A plywood model of the missing camera was added to complete the stereo configuration.
131. This number includes the 3 LANYARD, 12 ARGON, and 4 engineering (two radiometric, STARAD and one R&D) missions.

CHAPTER 4. POSTWAR STRATEGIC RECONNAISSANCE AND THE GENESIS OF CORONA

Note: I am indebted to numerous overhead reconnaissance pioneers who read and commented on the preliminary draft of this chapter and enriched it with their own recollections and documentary contributions. They are James Baker, James Coolbaugh, Merton Davies, Richard Garwin, William Hawkins, Jack Herther, William King, Richard Leghorn, Walter Levison, Frederic Oder, Bernard Schriever, Dow Smith, Robert Truax, William Troetschel, Paul Worthman, and Herbert York.

1. Strategic reconnaissance, to be sure, is a practice as old as warfare. What made it truly different after World War II was the acceptance of peacetime overflight and the appearance of the technical innovations (cameras, lenses, films, aircraft, and spacecraft) that made overflight reconnaissance possible. This form of reconnaissance, of course, embraced all sources, that is, signals intelligence (electronic, telemetry, and communications traffic) as well as imaging intelligence (radar and photography). This study focuses on the latter category because it directly answered the most pressing Cold War questions: Were the Soviets massing bombers along its northern shores, and was the U.S. the victim of a bomber and missile gap?
2. Donald E. Welzenbach, "Strategic Overhead Reconnaissance," unpublished draft manuscript, 1995, chapter 1.

3. George W. Goddard, "Photography Remains King in the Aerospace Age," *Photogrammetric Engineering* (March 1962): 88–89.
4. In addition to the K-22, during the war Baker designed a 60-inch f/6 telephoto lens—a folded optical system using mirrors; a 36-inch f/8 telephoto lens; a 4-inch f/2.8, spherical rotating lens that exposed spherical "shell" plates; and a 36-inch f/8 fluorite lens utilizing one element of synthetic optical fluorite that yielded "a perfection of color correction not hitherto achieved." At war's end, he was completing work on the 60-inch f/5 sealed telephoto lens that covered a 9- by 18-inch negative area. Considering these wartime contributions, Col. George Goddard judged Baker to be "the most versatile optical designer known to this command." Quotes from Col. George W. Goddard, Chief, Photographic Laboratory, Air Technical Service Command, letter to Gen. H. H. Arnold, Commanding General, Army Air Forces, July 17, 1945, as cited in Col. M. M. Irvine, War Dept Liaison Officer to National Defense Research Committee, letter to Office of Scientific Research and Development, Subject: "Transfer of Harvard University Records to Army Air Forces; Projects AC-29 and AC-88," April 17, 1946. See also Summary Technical Report of Division 16, NDRC, *Optical Instruments,* vol. 1 (Washington, D.C.: Government Printing Office, 1946), which contains details of the Harvard program. Baker wrote or coauthored a number of the chapters in this work.

   After World War II, James Baker worked as a consultant to Boston University's Optical Research Laboratory, Perkin-Elmer, and Eastman Kodak, while volunteering his services to committees of the USAF, the CIA, and PSAC. All of his postwar efforts had enormous ramifications for national security. Among other contributions, he designed and calibrated the high-resolution lenses for the 240-inch focal length "Boston Camera" as well as the lenses for the U-2, SR-71, and Eastman Kodak satellite cameras. Because almost all of Baker's government-related service involved classified projects, this modest man remained then, as he remains today, virtually unknown outside of the photogrammetry and intelligence communities.
5. "The U.S. Strategic Bombing Survey, Summary Report (European War)," September 30, 1945, as reprinted in *U.S. Strategic Bombing Surveys (European War and Pacific War)* (Washington, D.C.: Government Printing Office, October 1987), 39 (emphasis added), 41–42.
6. Welzenbach, "Strategic Overhead Reconnaissance," chap. 3.
7. George W. Goddard with DeWitt S. Copp, *Overview: A Lifelong Adventure in Aerial Photography* (Garden City, N.Y.: Doubleday, 1969), 349–50.
8. Back at the Harvard Yard, an unsatisfied Harvard University President James B. Conant insisted that the school not profit in any way from wartime military research. Despite a government offer of the Harvard Optical Laboratory to the school for the price of $1.00, Conant ordered the brand-new structure razed to the ground. The demolition, completed in June 1946, obliterated all signs of military optical research from campus, if not from memory.
9. Including representatives of Eastman Kodak, Polaroid, Bill Jack Optical, Fairchild Camera, Bausch & Lomb, Hycon, Perkin-Elmer, and Chicago Aerial.

10. Welzenbach, "Strategic Overhead Reconnaissance," chap. 3; Goddard, *Overview,* 351.
11. Richard S. Leghorn, "Objectives for Research and Development in Military Aerial Reconnaissance" (December 1946 [unpublished]), 26–27 (emphasis added). Beside Macdonald and Leghorn, the speakers included Maj. Gen. Curtis E. LeMay and Col. George W. Goddard, among others.
12. International Convention for Aerial Navigation, 1919, Article 1, as reprinted in De Forest Billyou, *Air Law,* 2d ed. (New York: Ad Press, 1964), 17.
13. Soviet aircraft from the Kola Peninsula could fly routes over the North Pole to attack the United States. By 1947, most American political and military leaders had come to view the atomic bomb as not just a larger, more destructive aerial bomb, but, if delivered in numbers, as a potentially decisive weapon. Gen. George C. Kenney, the first Commander-in-Chief, Strategic Air Command (CINCSAC), believed this and he directed the attention of the Strategic Air Command to the Arctic regions both "as a route of SAC bombers [and] as an avenue for a Soviet atomic strike upon the United States." John T. Farquhar, "A Need to Know: The Role of Air Force Reconnaissance in War Planning, 1945–1953," Ph.D. diss., Ohio State University, 1991, 75, 101–2. For the next few years, the CIA's strategic reconnaissance chief recalled, American leaders remained preoccupied with "the Soviet bomber force and the threat it posed to North America, as well as to Europe." Richard M. Bissell Jr., with Jonathan E. Lewis and Francis T. Pudlo, *Reflections of a Cold Warrior: From Yalta to the Bay of Pigs* (New Haven: Yale University Press, 1996), 92.
14. The National Security Act, signed by President Harry S. Truman on July 26, 1947, created the National Military Establishment and separate military departments of the Army, Navy, and Air Force.
15. Letter to Acting Chief of Eastern European Affairs, Department of State (Stevens), February 16, 1948, and AFOIR-CM to AAC/CC, Subject: "Violation of Soviet Frontier," n.d. (ca. Jan.–Feb. 1948), entry 214, TS Control and Cables Section General Files (July 45–Dec. 54), folder 2-900/2-999 (Feb. 1948); memorandum, Executive to the USAF DCS/O (Donnelly) and Multiple Addressees, Subject: "DCS/O Meeting, 14 May 1948," May 14, 1948, entry 214, TS Control and Cables Section General Files (July 45–Dec. 54), folder 2-1600/2-1699 (May 1948); memo for the Record, "To Brief Background Facts on Establishment of 40-Mile Limit on Reconnaissance Flights in the Pacific Area," n.d., entry 214, folder 2-3300/2-3399; and SECAF (Symington), letter (unsigned carbon copy) to Secretary of State (Marshall), n.d., entry 214, folder 2-1500/2-1599 (May 1948), all in RG 341, NARA, St. Louis (hereafter referred to as NARA-SL).
16. For the story of the detection, see Charles A. Ziegler and David Jacobson, *Spying without Spies: Origins of America's Secret Nuclear Surveillance System* (Westport, Conn.: Praeger, 1995).
17. In his letter to the Air Force Chief of Staff, LeMay observed: "Assuming that as a democracy we are not prepared to wage preventive war, this course of action poses two most different requirements: (1) An intelligence system which can locate the vulnerable elements of the Soviet striking force and forewarn us when attack by that force is imminent, and (2) Agreement at top governmental level that when

such information is received the Strategic Air Command will be directed to attack." Lt. Gen. Curtis E. LeMay, letter to Gen. Hoyt S. Vandenberg, December 12, 1949, as reprinted in Peter J. Roman, "Curtis LeMay and the Origins of NATO Atomic Targeting," *Journal of Strategic Studies* 16 (March 1993): 49. Although the first requirement would be adopted as national policy, the second was not. LeMay nonetheless remained a proponent of preemption, and in the years that followed seemingly took a perverse delight in explaining the concept and how he would execute it to civilian "policy experts" who visited his office at SAC Headquarters.

18. This effort eventually included Project Lincoln at MIT and the creation of the Distant Early Warning system, a radar picket line across Northern Alaska, Canada, Greenland, and Iceland. Eva C. Freeman, ed., *MIT Lincoln Laboratory: Technology in the National Interest* (Lexington, Mass.: MIT, 1995), 2–9; and George E. Valley Jr., "How the SAGE Development Began," *Annals of the History of Computing* 7 (July 1985): 196–226.
19. This review by State and Defense resulted in NSC-68, "United States Objectives and Programs for National Security," approved on September 30, 1950. See *Foreign Relations of the United States, 1950* (Washington, D.C.: Government Printing Office, 1977), 1:236–92, 400–401. The first paragraph of the conclusion declared: "Within the next four or five years the Soviet Union will possess the military capability of delivering a surprise atomic attack of such weight that the United States must have substantially increased general air, ground, and sea strength, atomic capabilities, and air and civilian defenses to deter war and to provide reasonable assurance, in the event of war, that it could survive the initial blow and go on to the eventual attainment of its objectives" (287–88).
20. In a cable to JCS Commands (today called Specified Commands), General of the Army and JCS Chairman Omar Bradley warned them that "the current situation in Korea has greatly increased the possibility of a general war," and directed that each "take such action as is feasible to increase readiness without creating atmosphere of alarm." JCS, cable to Lt. Gen. Curtis E. LeMay and other JCS Commanders, December 6, 1950, Box B-196, Papers of Curtis E. LeMay, LC. Truman had phoned Eisenhower on December 18, 1950, and asked him to return to active duty as SACEUR. Eisenhower, who believed in collective security and the NATO concept, accepted, and at month's end traveled to Washington to confer with government leaders. Afterward his son, John Eisenhower, recalled, "He expressed to me his disgust with the terrified atmosphere pervading all of Washington, from the President on down." John Eisenhower, *Strictly Personal* (Garden City, N.Y.: Doubleday, 1974), 156–57.
21. As late as October 5, 1950, the USAF Director of Intelligence, Maj. Gen. Charles P. Cabell, saw no hope of securing permission for overflights of the Soviet Union from the Departments of State, Defense, and the JCS. Responding to a SAC Headquarters request that the Air Force seek authorization to overfly the Kola Peninsula and the Chukotskiy Peninsula to determine "Soviet capabilities for delivering atomic bombs to targets within the United States," Cabell declined, adding, "If SAC wants formally to request it anyhow, I would recommend against

it, and unless SAC specifically requests otherwise, I would not forward it." He concluded, however, "[I am] looking forward to a day when it becomes either more essential or less objectionable." Communist Chinese intervention in Korea and positioning of Soviet forces in Eastern Europe a few weeks later clearly made these missions "more essential and less objectionable." Maj. Gen. Charles P. Cabell to Col. William A. Adams, SAC Director of Intelligence, October 5, 1950, TS Control and Cables Section General Files (July 45–Dec. 54), DCS/Operations, Director of Intelligence, entry 15621, RG 341, NARA-SL.

22. Farquhar, "A Need to Know," 142. For a listing of American aircraft lost to Soviet attacks, see Charles Maechling Jr., "Intrusions, Overflights, and Shootdowns," *Air Power History* 36 (Summer 1989): 6–15.
23. Gen. Nathan F. Twining, interview by John T. Mason Jr., August 17, 1967, in Arlington, Va. (third of four interviews), "Eisenhower Administration Project," 130–32, Oral History Research Collection, Butler Library, Columbia University; and Brig. Gen. Richard C. Neeley, USAF (Ret.), telephone interview by R. Cargill Hall, August 1, 1995. (Neeley was the pilot that SAC selected to fly this reconnaissance mission.) In his interview, Twining asserted that the Joint Chiefs wanted to use a new B-47 for this purpose and that President Truman signed papers approving the first overflights. Such papers, among the most closely held of the Cold War, have not yet been located. President Eisenhower later approved overflights verbally, but initialed the flight plans "DDE."
24. Maj. Gen. Carl A. Brandt, USAF Director of Requirements, DCS/Development, memorandum to Directorate of Intelligence, DCS/Operations, Subject: "Intelligence Requirement for B-47s for Special Reconnaissance Missions," January 4, 1951, with enclosure, "Memorandum for Record," DCS/O– Directorate Intelligence, TS Control and Cables Section General Files (July 45–Dec. 54), Box 2-17300 (1950) to 2-18299 (1951), folder 1-17300/2-17399, RG 341, NARA-SL. The memorandum identifies B-47B 49-2645 by tail number as the vehicle to be delivered on April 25, 1951, modified with special compass and autopilot equipment, and a high-latitude directional gyro system. For this first mission, identified in a classified addendum as "Project WIRAC," a special bomb-bay capsule had to be designed and fabricated to contain the cameras and associated equipment.
25. Neeley interview.
26. Records containing the terms and conditions of the British-American agreement have not been located and doubtless remain classified. Descriptions of the training and the missions flown have appeared in the memoirs and published recollections of RAF crew members. See, for instance, Squadron Leader John Crampton, RAF (Ret.) "The Royal Air Force RB-45C Special Duty Flight, 1951–1954," an address to the RAF Historical Society at the RAF Staff College, Bracknell, March 22, 1996, and Rex Saunders, letter to R. Cargill Hall, July 9, 1996.
27. Dino A. Brugioni, telephone interview by R. Cargill Hall, November 1, 1995. The tension and profound concern is evident in contemporary National Security Council deliberations publicly released. A 1952 national military evaluation determined that the United States would face "unavoidable defeat" if "a certain

number of targets in the U.S. were destroyed." Moreover, "the Soviet Union is capable of producing the requisite number of atomic, or thermonuclear, bombs to destroy those targets and is capable of producing the means of delivering the bombs." In August 1952 the CIA estimated that the USSR possessed a stockpile of thirty to fifty atomic weapons with an energy yield between thirty and seventy kilotons. The intelligence agency projected that the number of weapons would increase to 100 in mid-1953, 190 in mid-1954, and to 300 in mid-1955. *Foreign Relations of the United States, 1952–1954,* vol. 2: *National Security Affairs,* pt. 1 (Washington, D.C.: Government Printing Office, 1984), 14, 105, and 232, respectively.

28. Maj. Gen. Robert W. Burns, USAF Acting Deputy Chief of Staff, Operations, to Gen. Curtis E. LeMay, CINCSAC, Subject: "Special Aerial Photographic Operations," July 5, 1952, Accession 810-60, Package 129, Records of U.S. Air Force Commands, Activities, and Organizations, RG 342, NARA. For Tupolev's reverse engineering of the B-29, see Steven J. Zaloga, *Target America: The Soviet Union and the Strategic Arms Race, 1945–1964* (Novato, Calif.: Presidio, 1993), 63–79.
29. Robert A. Lovett, Secretary of Defense, memorandum to General of the Army Omar N. Bradley, Chairman, Joint Chiefs of Staff, Subject: "Reconnaissance Requirements," August 12, 1952, TS Accession 810-60, package 129, RG 342, NARA.
30. Maj. Gen. R. M. Ramey, USAF Director of Operations, to Gen. Curtis E. LeMay, CINCSAC, Subject: "Special Aerial Photographic Operations," August 15, 1952, with attachment: Lt. Col. P. O. Robertson, Acting Chief, Reconnaissance Division, memorandum to General Montgomery, Subject: "Project 52 AFR-18 [Instructions]," August 13, 1952, Accession 810-60, package 129, RG 342, NARA; and Col. Donald E. Hillman, USAF (Ret.), with R. Cargill Hall, "Overflight: Strategic Reconnaissance of the USSR," *Air Power History* 43 (Spring 1996): 28–39.
31. Headquarters USAF, Directorate of Intelligence, "Briefing for the Secretary of the Air Force, 23 June 1953" (script), and "Index to Charts," June 22, 1953 (script), p. 6, Accession 81-0325, Box 1, Case 14, RG 342, NARA. General LeMay awarded each member of the aircrews a Distinguished Flying Cross for this hazardous reconnaissance mission in lieu of a Silver Star, which he would have preferred to give. But the latter award required justification at USAF Headquarters, an action that would acquaint too many people with the reason for the award. In May 1953, LeMay struck an agreement with Air Force Vice Chief of Staff Gen. Nathan F. Twining to award DFCs or Air Medals to SAC reconnaissance aircrews operating overseas, who "performed special missions from bases in the United Kingdom." Gen. Curtis E. LeMay, CINCSAC, letter to Gen. N. F. Twining, Chief of Staff, USAF, November 17, 1955, Box 60, Twining File, LeMay Papers, LC.

   The secrecy surrounding these JCS-directed "special missions" was so tight that many senior Air Force leaders without a "need to know" in their entire career knew nothing more than rumor. One of them, Gen. Horace M. Wade, SAC Commander of the Eighth Air Force, later Commander-in-Chief of U.S. Air Forces in Europe, who retired as Vice Chief of Staff of the USAF, reflected on these years: "We were desperate. We were desperate for intelligence from the inside of Russia. I have a feeling, if the truth were really known, that there was a B-47 that was flown from Alaska across Russia and landed in Turkey. . . . I can't prove it, but I have a feeling that this was done." Gen. Horace M. Wade, USAF (Ret.), interview by Hugh N. Ahmann, Portland, Oregon, October 10–12, 1978, USAF Oral History Collection, p. 316, Air Force Historical Research Agency, Maxwell AFB, Alabama.

32. Gen. Jacob E. Smart, USAF (Ret.), letter to R. Cargill Hall, April 8, 1996. Smart, who in the mid-1950s served Far East Air Forces (FEAF) as director of operations, continued: At Headquarters FEAF, "we selected the optimum date and time of each mission based on a wide range of factors including sun-angle, weather, status of crew, aircraft and equipment, perceived activity in the target area, preparedness of supporting units—notably real-time intelligence gathering, air-sea rescue, etc.—all with care to avoid alerting friend, foe, or the media that something unusual was under way or planned."
33. Unknown to American intelligence until some years later, the BISON did not perform well and the Soviets built about 100 of these jet bombers, only enough to equip three wings. See Bill Gunston and Yefim Gordon, "The Extinct Bison," *Air International* 49 (October 1995): 222–29; 49 (November 1995): 275–79; and 49 (December 1995): 342–47.
34. For Eisenhower's views of the importance of this intelligence, see Christopher M. Andrew, *For the President's Eyes Only: Secret Intelligence and the American Presidency from Washington to Bush* (New York: HarperCollins, 1995), 199–201, 220–21; for the role of Killian and Land in 1954, see R. Cargill Hall, "The Eisenhower Administration and the Cold War," *Prologue: Quarterly of the National Archives* 27 (Spring 1995): 62–63.
35. Documents pertaining to the president's approval of this Soviet overflight have not been located but most likely reside in CIA, OSD, or JCS files instead of the Eisenhower Library. (It was the president's custom to listen to an overflight proposal and, if he approved it, initial the flight plan. In the case of the U-2, that document was returned to the CIA by Allen Dulles.) Learning of this and preceding overflights, Director of Naval Intelligence Rear Adm. Carl F. Espe requested photographs of the Soviet Union from the Air Force Director of Intelligence, Maj. Gen. J. A. Samford. Espe cited a total of six such flights, including the most recent one on May 8. Director of Naval Intelligence, Rear Admiral Carl F. Espe, memorandum to Director of Intelligence, USAF, Maj. Gen. J. A. Samford, Subject: "Photography, Request For," May 25, 1954, in entry 214, Box 77, Folder 4-1114/1129, RG 341, NARA. Samford's reply has not been found in Air Force or Navy archives or records.

Beginning with the Korean War, some historical evidence suggests that the JCS did delegate to JCS commanders, under certain circumstances, authority to conduct reconnaissance missions close to, or limited overflights of, the littoral regions of Communist China and the Soviet Union. See Gen. Bryce Poe II, "The Korean War: An Airman's Perception," draft paper cleared for public release on July 19, 1995, pp. 2, 23–25. Regrettably, when word of SAC-generated overflights of the Soviet Union became public knowledge before any pertinent Cold War records had been declassified, some authors in search of a conspiracy concluded that one JCS commander, CINCSAC General Curtis LeMay, authorized all of them without permission in a vain attempt to provoke the Soviet Union into starting World War III. Purposefully culling and arranging quotes from interviews, and ignoring all the conventions of scholarship, they affirmed for readers that Presidents Eisenhower and Kennedy had to contend with a real "General Ripper" loose in the national military establishment. See, for instance, Richard Rhodes, "The General and World War III," *New Yorker,* June 19, 1995, 47–59, and Paul Lashmar, "Killer on the Edge: The Warmongering Career of Curtis LeMay," *New Statesman and Society,* September 15, 1995, 20–22.

36. Crampton, "Royal Air Force RB-45C Special Duty Flight"; and James G. Baker, interview by R. Cargill Hall, May 9, 1996.
37. Eugene P. Kiefer, letter to Donald Welzenbach, March 16, 1988; Harold F. ("Bud") Wienberg, interview by R. Cargill Hall, March 16, 1995 (hereafter Wienberg interview); and Merton E. Davies and William R. Harris, *RAND's Role in the Evolution of Balloon and Satellite Observation Systems and Related U.S. Space Technology,* R-3692-RC (Santa Monica, Calif.: RAND Corporation, 1988), 33; see also Thomas A. Sturm, *The USAF Scientific Advisory Board: Its First Twenty Years, 1944–1964* (reprint: Washington, D.C.: Air Force History Office, Government Printing Office, 1986), 44–45, 48, and appendix C, "SAB Membership Roster." The Davies and Harris volume is the most comprehensive and thorough survey available on these subjects.
38. For an unclassified account of this project, later known as MOBY DICK and GENETRIX, see Curtis Peebles, *The Moby Dick Project: Reconnaissance Balloons Over Russia* (Washington D.C.: Smithsonian Institution Press, 1991).
39. Camera shutter and lens technology, they knew, would require major improvement to move up from the altitude of 13 miles, employed for balloon and aerial platforms, to an altitude of 300 miles then planned at RAND for a reconnaissance satellite and still provide useful images of objects on the earth's surface.
40. Davies and Harris, *RAND's Role,* 35–38.
41. The contents of the AFDAP intelligence and reconnaissance DPO was described by Richard S. Leghorn in comments on an early draft of this study, July 5, 1995. A copy of the I&R DPO has not been found.
42. Jay Miller, *Lockheed U-2* (Arlington, Tex.: Aerofax, 1983), 10–12, 17–18. Miller's account is based on documents provided by and an interview by a principal at WADC, Maj. John Seaberg. The X-16 effort was canceled shortly after the U-2 began test flights in August 1955.
43. Welzenbach, "Strategic Overhead Reconnaissance," chap. 8. For other authoritative accounts, see Miller, *Lockheed U-2;* also Clarence L. "Kelly" Johnson with Maggie Smith, *Kelly: More than My Share of It All* (Washington, D.C.: Smithsonian Institution Press, 1985), chap. 13; and Ben R. Rich with Leo Janos, *Skunk Works: A Personal Memoir of My Years at Lockheed* (New York: Little, Brown, 1994), chaps. 6 and 7.
44. Burton Klein, telephone interview by R. Cargill Hall, October 4, 1995.

45. Ibid.
46. Not only would this unarmed single-engine aircraft fly higher, but, as Killian later told Herbert York, it would be "manifestly less hostile" than Air Force reconnaissance bombers used on overflight missions. Herbert F. York and G. Allen Greb, "Strategic Reconnaissance," *Bulletin of the Atomic Scientists* 33, no. 4 (April 1977): 35. Also, regarding U-2 origins, James G. Baker, letter to R. Cargill Hall, December 21, 1993. Baker was a member of Din Land's TCP intelligence committee and designed the remarkable B-2 camera later employed in the U-2. He also served as chairman of the Air Force Scientific Advisory Board's intelligence systems panel in 1954. As he recollected events, Allen Donovan brought word of Kelly Johnson's CL-282 to members of the Air Force panel at a meeting after Baker had returned from a trip to Europe in early March (the meeting would have had to follow in late March–July). "We kept these discussions very close indeed and carried them over into the TCP not long afterward. As a result, Din met at Lockheed with Kelly Johnson and called me from California. Din said words I cannot forget: "'Jim, I think we have your airplane.'"
47. A. J. Goodpaster, "Memorandum of Conference with the President, 0810, Nov. 24, 1954," "ACW Diary, Nov 1954," A7, Bay 3, Anne C. Whitman Series, Anne Whitman File, Dwight D. Eisenhower Library, Abilene, Kansas (hereafter DDE). The best accounting of events leading to the November 24 meeting appear in Dino A. Brugioni, *Eyeball to Eyeball: The Inside Story of the Cuban Missile Crisis* (New York: Random House, 1990), 16–19; James R. Killian Jr., *Sputnik, Scientists, and Eisenhower: A Memoir of the First Special Assistant to the President for Science and Technology* (Cambridge, Mass.: MIT Press, 1977), 82; and Michael R. Beschloss, *Mayday: The U-2 Affair* (New York: Harper and Row, 1991), 82; also Donald Welzenbach, "Din Land: Patriot from Polaroid," *Optics and Photonics News* 5 (October 1994): 23–24.
48. A biographical sketch of the versatile and complex Richard Bissell appears in Evan Thomas, *The Very Best Men* (New York: Simon and Schuster, 1995). For his own posthumous accounting, see Bissell, *Reflections of a Cold Warrior.*

    For an account of the significant intelligence results of the U-2 missions, see the statement of Allen W. Dulles, DCI, before the U.S. Senate Committee on Foreign Relations on May 31, 1960, regarding "Events Incident to the Summit Conference," in U.S. Senate, *Executive Sessions of the Senate Foreign Relations Committee,* Historical Series, vol. 12, 86th Congr., 2d sess., 1960 (declassified and made public November 1982), 280–87.
49. Ritland and Mixson chose U-2 pilots from an Air Force pool. Once chosen, they were seconded to the CIA from the Air Force. They were restored to military status after a period of service with the CIA. Wienberg interview. Although CL-282 and the U-2 shared a basic design concept, the latter aircraft differed substantially in configuration and equipment from the original Lockheed proposal. For example, a Pratt & Whitney J57 axial flow turbojet engine recommended by John Seaberg powered the U-2 instead of the GE J73 engine that Kelly Johnson first specified.

50. In 1955 Nelson Rockefeller served as special assistant to the president for psychological warfare, while Harold Stassen served as special assistant to the president for disarmament.
51. Gen. Andrew J. Goodpaster, USA (Ret.) letter to R. Cargill Hall, January 11, 1996; and John Eisenhower, letter to R. Cargill Hall, January 9, 1996. According to Leghorn, Stephen Posony, another member of Schriever's AFDAP team from Air Force intelligence, helped draft the Open Skies proposal for Nelson Rockefeller. Solid accounts appear in Brugioni, *Eyeball to Eyeball,* 25–26; Eisenhower, *Strictly Personal,* 177–78; and Dwight D. Eisenhower, *The White House Years: Mandate for Change* (Garden City, N.Y.: Doubleday, 1963), 519.
52. "Statement on Disarmament, July 21," *Department of State Bulletin* 33, no. 841 (August 1, 1955): 174. The term "Open Skies" was coined later by the popular press and applied to this disarmament plan. The background of this proposal, as debated in the National Security Council, is contained in *Foreign Relations of the United States, 1955–1957,* vol. 20, *Regulation of Armaments; Atomic Energy* (Washington, D.C.: Government Printing Office, 1990), see esp. docs. 33 through 48.
53. The most authoritative account of these events appears in W. W. Rostow, *Open Skies: Eisenhower's Proposal of July 21, 1955* (Austin: University of Texas Press, 1982).
54. NSC 5522, "[Executive Department] Comments on the Report to the President by the Technological Capabilities Panel of the Science Advisory Committee," June 8, 1955. For instance, see comments by Donald Quarles for the Department of Defense and Allen Dulles for the CIA at pp. A15-A44, and A45-A56. White House Office, Office of the Special Assistant for National Security Affairs: Records, 1952–61, NSC Policy Papers, Box 16, Folder NSC 5522 Technological Capabilities Panel, DDE. Also, R. Cargill Hall, "The Eisenhower Administration and the Cold War: Framing American Astronautics to Serve National Security," *Prologue* 27 (Spring 1995): 59–72; and Hall, "The Origins of U.S. Space Policy: Eisenhower, Open Skies, and Freedom of Space," in *Exploring the Unknown: Selected Documents in the History of the U.S. Civil Space Program,* vol. 1, ed. John Logsdon et al., NASA SP-4407 (Washington, D.C.: Government Printing Office, 1995), 225–33.
55. Richard S. Leghorn, interview by R. Cargill Hall and Donald Welzenbach, December 13, 1995. Leghorn's views of strategic reconnaissance employed as an arms control and disarmament inspection system at this time appeared in a seminal article: "U.S. Can Photograph Russia from the Air Now: Planes Available, Equipment on Hand, Techniques Set," *U.S. News and World Report,* August 5, 1955, 70–75.
56. Paul E. Worthman recollections cited by Rostow in *Open Skies,* 189–94; Tom D. Crouch, *The Eagle Aloft: Two Centuries of the Balloon in America* (Washington, D.C.: Smithsonian Institution Press, 1983), 644–49; and Peebles, *The Moby Dick Project.*
57. Levison, interview by R. Cargill Hall, May 9, 1996 (hereafter Levison interview); Levison, letter to Hall, September 2, 1996.

58. Crouch, *Eagle Aloft,* 644–49; and Levison interview. In the event aerial retrieval failed, the camera-carrying gondolas were designed to float on the ocean's surface and radiate a signal for 24 hours before a plug dissolved and the gondola sank.
59. Bissell, *Reflections of a Cold Warrior,* 112.
60. Sergei Khrushchev, interview by R. Cargill Hall and Richard S. Leghorn, Providence, R.I., July 5, 1995.
61. See Hall, "Eisenhower Administration and the Cold War."
62. Executive Order 10656, February 6, 1956, as cited in *U.S. Government Organization Manual, 1958–1959* (Washington, D.C.: Government Printing Office, 1958), 538.
63. Kenneth E. Greer, "CORONA," *Studies in Intelligence,* Supplement 17 (Spring 1973), as reprinted in *CORONA: America's First Satellite Program,* ed. Kevin C. Ruffner, Center for the Study of Intelligence (Washington, D.C.: Government Printing Office, 1995), 4–5.
64. Jacob Neufeld, *Ballistic Missiles in the United States Air Force, 1945–1960* (Washington D.C.: Office of Air Force History, 1990). The Teapot Committee report and recommendation, as it was popularly known, is reprinted at appendix 1 of the Neufeld book.
65. William G. King, USAF (Ret.), telephone interview by R. Cargill Hall, September 9, 1996; James S. Coolbaugh, "The Beginnings of the Air Force Satellite Program: A Memoir," contained in Space Policy Institute Collection, George Washington University; and Hall, "Origins of U.S. Space Policy," 218–21.
66. Troetschel, "An Early History of the Air Force Space Program," n.d., contained in SPI; Coolbaugh, "Beginnings of the Air Force Satellite Program"; and King interview, September 9, 1996.
67. Back in the spring of 1956, Truax and Coolbaugh "flew a B-25 up and down the West Coast looking for the best spot to locate a satellite launch facility. RAND's recommendation of Alaska as the place to locate such a facility was long forgotten. We finally settled on two sites. The ideal place for polar launches was the south side of the Army's Camp Cooke, which was located on Point Arguello, about 55 miles WNW of Santa Barbara. The other site which could have been used was near Santa Cruz. (This latter location became Lockheed's Santa Cruz Test Facility, where the Agena stage was 'hot'-fired as part of its pre-launch validation.) We were lucky in our selection of the Camp Cooke site because the Navy had a radar tracking site there and Bob knew the Navy officer in charge of the operation, Commander Bob Frietag. When 'the dust settled,' the Air Force was authorized to build a launch base for satellites there. This action predated the Air Force's acquisition of Camp Cooke for a missile launch base by about nine months." Coolbaugh, "Beginning of the Air Force Satellite Program," 44.
68. Hall, "Origins of U.S. Space Policy," 224.
69. Amrom H. Katz and Merton E. Davies, "On the Utility of Very Large Satellite Payloads for Reconnaissance," RAND D-5817, November 14, 1958, 6.
70. Davies and Harris, *RAND's Role,* 69–70. A number of conditions prefigured the choice of a long-lived readout reconnaissance satellite. Beside the demand for indications and warning of surprise attack that arose in the 1950s, when RAND

conducted its early satellite studies, returning anything from earth orbit was judged technically unfeasible. Moreover, without an ICBM available, a satellite program also would have to develop and pay for its own booster. Finally, without the re-entry option, whatever was placed in orbit would have to function for a long time, at least one year, to justify the expense of getting it up there. For a discussion of these conditions, see Amrom H. Katz, "Some Notes on the Evolution of RAND's Thinking on Reconnaissance Satellites," RAND D-4753, November 27, 1957.

Studies of a recoverable film payload and ways to protect it from the searing heat of atmospheric reentry that began in 1956 continued at RAND throughout the year. See J. H. Huntzicker and H. A. Lieske, "Physical Recovery of Satellite Payloads—A Preliminary Investigation," RAND RM-1811, June 26, 1956. The authors estimated that a payload of 50 pounds of film could be recovered from a satellite weighing about 225 pounds.

71. Merton E. Davies, telephone interview by R. Cargill Hall, July 2, 1996. Davies eventually received a U.S. patent for his "spin-pan camera."
72. Ibid., 86–87; and M. E. Davies and A. H. Katz et al., "A Family of Recoverable Reconnaissance Satellites," RAND RM-2012, November 12, 1957.
73. With the declassification of many Eisenhower administration records, the covert IGY satellite policy, first surmised by Walter MacDougall and Stephen Ambrose, is absolutely confirmed. (See chapter 5.) The recommendations for it outlined in James Killian's TCP report of February 1955 are contained in NSC 5522, "Comments on the Report to the President." See, for instance, those sections submitted by the CIA and Departments of State and Defense. The NSC 5522 report is in White House Office, Office of the Special Assistant for National Security Affairs: Records, 1952–1961, NSC Policy Papers, Box 16, Folder NSC 5522 Technological Capabilities Panel, DDE. See also Hall, "Eisenhower Administration and the Cold War," 59–72; and Dwayne A. Day, "A Strategy for Space: Donald Quarles, the CIA and the US Scientific Satellite Programme," *Spaceflight* 38 (September 1996): 308–12.
74. Greer, "CORONA," 4–5.
75. Ibid., 5. According to Greer, no records of this briefing were kept and the outcome is surmised from subsequent decisions.
76. Davies and Katz, "Family of Recoverable Reconnaissance Satellites."
77. Col. F. C. E. Oder, WS-117L Director, memorandum to Maj. Gen. B. A. Schriever, Commander, AFBMD, no subject, August 27, 1957; and Lockheed Missiles and Space Division, WS-117L Development Plan for Program Acceleration, LMSD-2832, January 6, 1958.
78. Richard E. Horner, Assistant Secretary of the Air Force (R&D), memorandum to Neil McElroy, Secretary of Defense, Subject: "Outer Space Vehicles," November 12, 1957.
79. Col. F. C. E. Oder, WS-117L director, memorandum to Maj. Gen. B. A. Schriever, no subject, December 7, 1957. After the meeting in Washington, Schriever returned to the West Coast and met with Air Force program participants on December 5. They included Robert Gross and Willis Hawkins of Lockheed, as well as Colonel Oder and General Schriever from AFBMD. They agreed on a

configuration of the film-recovery satellite that used a Thor booster and Lockheed Agena upper stage equipped with horizon sensors and stabilized in space in a position horizontal to the earth. The Fairchild camera was to be mounted in a reentry capsule fixed to the Agena's forward end. Mounted on a drive shaft, the capsule was to be spun up and released from the Agena after the ensemble had stabilized on orbit. This approach appears in Lockheed's development plan released a few weeks later. WS-117L Development Plan for Program Acceleration, LMSD-2832, January 6, 1958. Both Oder and Hawkins have affirmed that by January 1958 the AFBMD and Lockheed participants intended to abandon this separable spinning payload and fix a variant of the Fairchild camera directly to the attitude-stabilized Agena. Though doubtless true, this latter plan does not appear in any contemporary documents thus far located. It was unquestionably adopted in February and March 1958, using another, different kind of camera. However, James Plummer has stated that the initial CORONA design he was told to pursue was Davies and Katz's spinner with the Aerobee upper stage, not a vehicle using the Agena.

80. Brig. Gen. A. J. Goodpaster, "Memorandum of Conference with the President," (meeting in Oval Office on February 7 with Land, Killian, and Goodpaster), February 10, 1958, White House Office of Staff Secretary, Alpha Series, Box 14, Intelligence Matters, DDE.
81. Greer, "CORONA," 6.
82. Ibid., 5.
83. Roy W. Johnson, Director, ARPA, memorandum to James H. Douglas Jr., Secretary of the Air Force, Subject: "Reconnaissance Satellites and Manned Space Exploration," February 28, 1958.
84. Herbert F. York, letter to R. Cargill Hall, August 24, 1996. York said, "Thereafter, I frequently testified before the Congress and gave press statements about how Discoverer was a great engineering program for the development of space maneuver, recovery, life support, etc. . . . I had to face the wrath of RAND . . . (Katz, Buchheim, etc.) and explain how we and others had reviewed their recoverable satellite ideas and decided to place our bets on the [Air Force] 'readout' system instead." Also York, letter to Hall, September 13, 1996. Quarles, York added, "was very influential, but he was also very low key and soft-spoken, so history largely ignores him. He was widely thought to be conservative and unimaginative. He was conservative in approach, but he was smart and much more knowledgeable about technical issues than anyone else in the Pentagon before the changes wrought by Sputnik. In brief, he quickly understood whatever was brought to his attention and he was decisive."

To be sure, not everyone perceived space reconnaissance favorably. Late in February or early March 1958, Robert Truax recalled, ARPA director Roy Johnson briefed the Joint Chiefs of Staff and mentioned the Air Force WS-117L, which he described as " 'the most important weapon system under development in the country.' I got a hurry-up call to go brief '21 knot' [Admiral Arleigh] Burke, the Chief of Naval Operations, who didn't know what WS-117L was. After my briefing he snorted: 'Why, it is nothing but a reconnaissance system! If it doesn't hit the other guy with something hard, it can't be important.'" Robert C. Truax, letter to R. Cargill Hall, September 12, 1996.

85. Wienberg interview. For Bissell and Ritland, Katz's public reaction made credible for the knowledgeable world at large that the film-recovery project had indeed ended, when in fact it had not. A year or so later, however, as long-time comrades ceased speaking with them about space reconnaissance, Davies and Katz realized that the project had continued and that they would not be asked to contribute. The bitter aftertaste of that knowledge would remain with them for many years.
86. No minutes were taken and years later attendees produced two accounts of the naming. In the first version, when Bissell looked around the room and asked, "What shall we call this project?" a participant reportedly removed the paper ring from his cigar and said, "Why not CORONA?" In the second version, a participant pointed to a typewriter on a desk nearby and said, "Why not CORONA?" Whichever version is correct, the name stuck.
87. Leghorn, interview by Welzenbach and Hall, December 13, 1995.
88. Levison interview, and Richard W. Philbrick, interview by R. Cargill Hall, May 10, 1996; Amrom Katz, letters to Walter Levison, March 14, 1957, and January 3, 1958. A description and diagram of the HYAC camera appears in Davies and Harris, *RAND's Role,* 78–85.
89. Davies and Harris, *RAND's Role,* 29–30; When Duncan Macdonald became graduate dean, F. Dow Smith succeeded him as chairman of the BU Physics Department, as well as director of the BU Physical Research Laboratories. Smith encouraged Case to dispose of the BUPRL as a unit, and not to break the organization into pieces, as some alternate scenarios had proposed. F. Dow Smith, interview by R. Cargill Hall, April 22, 1996.
90. No records of this meeting have been located and accounts of surviving participants differ on the location. One source recalled the CORONA evaluators met in Cambridge, Mass., for the briefings. Jack Herther, who presented the Itek briefing with Duncan Macdonald, is certain it was held in the Old Executive Office Building because he remembers purchasing the airline tickets for the flight from Boston to Washington, the names of many of those present at the briefing, and that there was no contact with competitors (other contractors) because of the physical separation of the briefing teams. The results of this meeting set the stage for the subsequent decision to make Itek's HYAC the primary instead of the backup camera.
91. For an authoritative description of Itek's HYAC camera scaled for space flight, see F. Dow Smith, "The Eyes of CORONA: The World's First Satellite Reconnaissance Program," *Optics and Photonics News* 6, no. 10 (October 1995): 34–39. To improve resolution at the earth's surface, Itek eventually combined a faster high-resolution Petzval lens and fine-grain film in place of focal length. That design tradeoff belied George Goddard's maxim for close-up photography: "There is no substitute for focal length." Or, as rephrased by Amrom Katz: "If you want close-up pictures, get close up!"

    Goddard, who urged greater focal length in terms of feet, not inches, likely would have disapproved of the Itek design compromise. Perhaps the greatest expression of his approach was the immense "Boston Camera" constructed at BU's Optical Research Laboratory. It featured a James Baker-designed 240-inch focal length f/8 lens (with the forward lens, made of a single blank of Schott glass, 32 inches in diameter), and it produced pictures on an 18- by 36-inch film format, with a CR-39 plastic filter chemically designed to achieve a coating that duplicated Eastman Kodak's Wratten 21. Completed and delivered to the Air Force in 1951, this camera was mounted in the largest available transport, the double-decker Boeing C-97, and used in the LOROP (long-range oblique photography) program that employed aircraft cameras to look across denied borders. Retired in the 1960s, the camera is now on display at the USAF Museum at Wright-Patterson AFB in Ohio. F. Dow Smith, letter to R. Cargill Hall, September 3, 1996, and Baker, interview by Hall, May 9, 1996.

92. John C. Herther and Malcolm R. Malcomson, "A Transition Control System," submitted in partial fulfillment of the requirement for the master of science degree at MIT, May 23, 1955; and John C. Herther, interview by R. Cargill Hall, July 27, 1996 (hereafter Herther interview).
93. Herther interview. To be sure, the Lockheed Missile and Space Division team directed by Jack Carter and James Plummer ultimately designed and fabricated Agena's ascent guidance and on-orbit stabilization system. At Itek, Herther served as the system integration engineer and also directed the flight environmental qualification testing of the CORONA camera, including vibration, shock, and the extended simulated orbital vacuum operation for diagnosing film breakage that occurred in space operations. Ultimately, the only fix for the breakage problem was to have Eastman Kodak replace 6-mil acetate-based film with the 2-mil polyester-based film. That exchange yielded the extra benefit of three times more pictures per pound! Herther subsequently became the Itek project manager for LANYARD, a CIA/Air Force CORONA follow-on involving a 66-inch focal length f/5 panoramic camera that achieved 2-foot resolution at the earth's surface on its single successful development flight in August 1963.
94. The unsolicited Itek camera proposal in February 1958 had referenced "unpublished correspondence" between Duncan Macdonald and Arthur Lundahl that correlated photo-interpretation experiments performed at BU and the CIA. These experiments confirmed that positive recognition of objects in photographs required a ground resolution size significantly smaller (by a factor of 3-to-5 depending on conditions) than the object size.
95. John C. Herther, letter to R. Cargill Hall, May 13, 1996.
96. Greer, "CORONA," 7–9; Herther interview.
97. "Project CORONA Outline," COR-0013, with cover letter from Richard M. Bissell Jr., to Col. Andrew Goodpaster, USA, COR-0014, both dated April 15, 1958.
98. No formal record of the April 16 meeting was kept. Beside a verbal approval, legend has it that Eisenhower scribbled "Okay, DDE," on the proposal, or, variously, Cabell noted his approval on the back of an envelope. Whatever the case, this putative "written" authorization has not yet been located.
99. CORONA Statement of Work, April 25, 1958.

CHAPTER 5. A STRATEGY FOR RECONNAISSANCE

1. Several authors, working independently and mostly from unclassified sources, reached this conclusion in the mid-1980s. Due to recent declassifications at several U.S. archives (most of the documents cited here were only declassified in 1995–96), it can now be proven beyond the shadow of a doubt that the American government was pursuing a definite strategy in its plan to launch a scientific satellite vehicle as part of U.S. participation in the International Geophysical Year (IGY). See Walter A. McDougall, *The Heavens and the Earth* (New York: Basic Books, 1985), and R. Cargill Hall, "Origins of U.S. Space Policy: Eisenhower, Open Skies, and Freedom of Space," in *Exploring the Unknown: Selected Documents in the History of the U.S. Civil Space Program,* vol. 1, ed. John M. Logsdon et al., NASA SP-4407 (Washington, D.C.: Government Printing Office, 1995). The author is indebted to their trail-blazing work.
2. J. R. Killian Jr., to Gen. Curtis E. LeMay, September 2, 1954, Papers of Curtis LeMay, Box 205, Folder B-39356, Manuscript Division, Library of Congress.
3. Information on the classified annexes comes from an interview by Donald E. Welzenbach with James Killian and is referenced in Donald E. Welzenbach, "Science and Technology: Origins of a Directorate," *Studies in Intelligence* 30 (Summer 1986), found in RG 263, NARA. Although the intelligence section of the TCP report remains classified and awaits declassification review, the index has been declassified. It includes the word "satellites," but apparently in the context of satellite countries of the USSR. "The Report to the President by the Technological Capabilities Panel of the Science Advisory Committee," February 14, 1955, Office of the Staff Secretary: Records of Paul T. Carroll, Andrew J. Goodpaster, L. Arthur Minnich, and Christopher H. Russell, 1952–61, Subject Series (hereafter OSS-SS), Alphabetical Subseries, Box 16, "Killian Report-Technological Capabilities Panel (2)," Dwight D. Eisenhower Library (hereafter DDE).
4. Gen. Andrew Goodpaster, interview by Dwayne A. Day, March 19, 1996 (hereafter Goodpaster interview). Goodpaster went to the White House in October 1954 as a colonel and was promoted to brigadier general while there. He eventually rose to the rank of general and assumed command of Supreme Headquarters Allied Powers Europe (SHAPE) in 1969.
5. Policy Planning Staff, Department of State, "Report to the President on the Threat of Surprise Attack," March 14, 1955, Box 87, "NSC 5522 Memoranda," General Records of the Department of State: Records Relating to State Department Participation in the Operations Coordinating Board and the National Security Council, 1947–1963, RG 59, NARA (hereafter General Records).
6. Robert R. Bowie, memorandum for Mr. Phleger, Policy Planning Staff, Department of State, March 28, 1955, Department of State Central Files, 711.5/3-2855.
7. Joseph Kaplan, Chairman, U.S. National Committee, International Geophysical Year 1957–58, National Academy of Sciences, to Dr. A. T. Waterman, Director, National Science Foundation, March 14, 1955.

8. Alan T. Waterman, Director, memorandum for Mr. Robert Murphy, Deputy Undersecretary of State, March 18, 1955.
9. Robert Murphy, memorandum for Dr. Alan T. Waterman, Director, National Science Foundation, April 27, 1955.
10. Alan T. Waterman, Director, to Donald A. Quarles, Assistant Secretary of Defense (Research and Development), May 13, 1955.
11. The newly released part of the document is in italics. NSC 5520, May 20, 1955, Box 112, "NSC 5520," General Records.
12. Ibid. This portion of the document remained classified until 1995.
13. James S. Lay Jr., Executive Secretary, National Security Council, Memorandum for the National Security Council, "U.S. Scientific Satellite Program," November 9, 1956, Box 86, "NSC 5520—US Scientific Satellite Program (Memoranda)," General Records.
14. Memorandum, "Discussion at the 250th Meeting of the National Security Council, Thursday, 26 May 1955," May 27, 1955, Files of Dwight D. Eisenhower as President, Ann Whitman File, NSC Series, Box 6, "250th Meeting of NSC, May 26, 1955," DDE.
15. National Security Council, NSC 5522, June 8, 1955, Comments on the Report to the President by the Technological Capabilities Panel, p. S-5, White House Office, Office of the Special Assistant for National Security Affairs: Records, 1952–61, NSC Policy Papers, Box 16, Folder NSC 5522 Technological Capabilities Panel, DDE.
16. Ibid., A-55-6
17. Robert R. Bowie, Policy Planning Staff, Department of State, "Recommendations in the Report to the President by the Technological Capabilities Panel of the Science Advisory Committee, ODM (Killian Committee): Item 2—NSC Agenda 10/4/56," Box 87, "NSC 5522 Memoranda," General Records.
18. Constance McLaughlin Green and Milton Lomask, *Vanguard: A History* (Washington, D.C.: Smithsonian Institution Press, 1971), 34–56.
19. Ibid., vi. Green and Lomask added: "To these observations, I can add from my own experience that inter-service rivalry exerted strong influence; also, that any conclusion drawn would be incomplete without taking into account the antagonism still existing toward von Braun and his co-workers because of their service on the German side of World War II."
20. Col. A. J. Goodpaster, "Memorandum for Record," June 7, 1956, White House Office, Office of the Staff Secretary: Records, 1952–1961, Box 6, "Missiles and Satellites," DDE (hereafter WHO-DDE).
21. Homer J. Stewart, Chairman, Advisory Group on Special Capabilities, memorandum for the Assistant Secretary of Defense (R&D), "VANGUARD and REDSTONE," June 22, 1956, WHO-DDE.
22. Stewart's memorandum was stamped "SECRET." There is some reason to believe it was actually written in May 1956 instead of June. It is rare for a report of a meeting to be written two months after the meeting. Furthermore, the memo also mentions the group's *upcoming* meeting on June 19 and 20 concerning the propulsion systems for Vanguard and invites contractor representatives to attend

this meeting, which would already have happened by the time the memo was written. The June 22 date may be a typographical error.

23. E. V. Murphree, Special Assistant for Guided Missiles, memorandum for Deputy Secretary of Defense, "Use of the JUPITER Re-entry Test Vehicle as a Satellite," July 5, 1956, WHO-DDE (emphasis added).
24. In another one-page memorandum from C. C. Furnas to the Deputy Secretary of Defense, dated July 10, 1956, and stamped "Secret," Furnas mentioned the meeting that he and Murphree had with Robertson on July 9. Furnas used this memorandum as a cover letter to forward the previous report to him by Homer Stewart's Advisory Group on Special Capabilities. He concluded by saying, "I trust that this will serve your purpose in reporting your evaluation of the suggestion that a Redstone vehicle will be used." C. C. Furnas, Assistant Secretary of Defense for Research and Development, memorandum for Deputy Secretary of Defense, July 10, 1956, WHO-DDE.
25. William Ewald, *Eisenhower the President: Crucial Days, 1951–1960* (Englewood Cliffs, N.J.: Prentice-Hall, 1981), 284.
26. Alan T. Waterman, National Science Foundation, memorandum to Mr. Percival Brundage, Bureau of the Budget, "Funding of Earth Satellite Program, International Geophysical Year," April 7, 1956, Box 86, "NSC 5520—US Scientific Satellite Program (Memoranda)," General Records.
27. Lay, memorandum for the National Security Council, "U.S. Scientific Satellite Program," November 9, 1956, with attached: "Draft Report on NSC 5520, U.S. Scientific Satellite Program Background."
28. The subject was also mentioned in a number of RAND documents dating as early as the late 1940s.
29. Goodpaster interview.
30. David Z. Beckler, Executive Officer, Science Advisory Committee, memorandum for General Goodpaster, Special Briefing, September 19, 1957, OSS-SS, Alphabetical Subseries, Box 23, "Science Advisory Committee (2) Sept.–Oct. 1957," DDE.
31. Brig. Gen. Andrew Goodpaster, memorandum of Conference with the President, October 7, 1957, OSS-SS, Department of Defense Subseries, Box 6, "Missiles and Satellites," DDE.
32. Brig. Gen. Andrew Goodpaster, memorandum of Conference with the President (following McElroy swearing in), October 9, 1957, OSS-SS, Department of Defense Subseries, Box 6, "Missiles and Satellites," DDE.
33. Waterman to Quarles, May 13, 1955.
34. Percival Brundage, Director, Bureau of the Budget, memorandum for the President, "Project VANGUARD," April 30, 1957, WHO-DDE.
35. Former CIA Deputy Director of Science and Technology Albert "Bud" Wheelon has speculated in conversations with the author that the money probably came from DCI Allen Dulles's substantial discretionary budget.
36. Goodpaster interview.
37. Eisenhower's comments on this subject appear in numerous documents. For instance, in October 1957 Goodpaster reported, "The President went on to say he

sometimes wondered whether there should not be a fourth service established to handle the whole missiles activity." Brig. Gen. A. J. Goodpaster, "Memorandum of Conference with the President, 11 October 1957, 8:30 A.M.," October 11, 1957, Ann Whitman File, DDE Diary Series, Box 67, "Oct. 57 Staff Notes (2)," DDE. In January 1958 Goodpaster reported, "In the course of the discussion the President indicated strongly that he thinks future missiles should be brought into a central organization." Brig. Gen. A. J. Goodpaster, "Memorandum of Conference with the President, 21 January 1958," January 22, 1958, OSS-SS, Department of Defense Subseries, Box 6, "Missiles and Satellites, Vol. 2 (1) [January–February 1958]," DDE. In February 1958, Goodpaster reported, "The President said that he has come to regret deeply that the missile program was not set up in OSD rather than in any of the services." Brig. Gen. A. J. Goodpaster, "Memorandum of Conference with the President, 4 February 1958 (following Legislative Leaders meeting)," February 6, 1958, OSS-SS, Department of Defense Subseries, Box 6, "Missiles and Satellites, Vol. 2 (1) [January–February 1958]," DDE.

38. See n. 37.
39. Robert Frank Futrell, *Ideas, Concepts, Doctrine* (Washington, D.C.: Government Printing Office, 1989), 1:589. The comments on Quarles's partisanship come from the interview by Goodpaster.
40. U.S. House of Representatives, *Organization and Management of Missile Programs, Hearings before a Subcommittee of the Committee on Government Operations,* 86th Cong., 2d Sess. (Washington, D.C.: Government Printing Office, 1959), 133.
41. Ibid.
42. McDougall, *The Heavens and the Earth,* 182–83.
43. Gordon Gray, Special Assistant to the President, to Brig. Gen. Andrew J. Goodpaster, February 16, 1959, with attached: James R. Killian Jr., to Gordon Gray, Special Assistant to the President, February 13, 1959; James R. Killian Jr., to Dr. T. Keith Glennan, Administrator, National Aeronautics and Space Administration, February 13, 1959; Lloyd V. Berkner, Chairman, Space Science Board, National Academy of Sciences, to Dr. James R. Killian Jr., Special Assistant to the President for Science and Technology, January 28, 1959; O. G. Villard Jr., Space Science Board, to Dr. L. V. Berkner, President, Associated Universities, Inc., January 22, 1959, OSS-SS, Department of Defense Subseries, Box 15, "Space [January–June 1959]," DDE.
44. Richard S. Leghorn, Political Action and Satellite Reconnaissance (Draft), April 24, 1959, OSS-SS, Department of Defense Subseries, Box 15, "Space [January–June 1959]," DDE.
45. The U-2 also raised once again the issue of where airspace ended and space began. At the Eleventh International Astronautical Federation Congress in Stockholm, Sweden, Spencer M. Beresford presented a paper that connected the U-2 and violations of international airspace with the possibility of future flights by military MIDAS and SAMOS spacecraft. A State Department official obtained a copy of Beresford's paper before his presentation and notified the U.S. Information Service in Stockholm that the paper raised a number of highly sensitive topics about which

the U.S. government should not comment. W. E. Gathright, to USIS-Stockholm, TOUSI II, Joint State USIA Message, August 12, 1960, with attached: Remarks of Spencer M. Beresford, United States of America, at the Eleventh Annual Congress of the International Astronautical Federation, Stockholm, Sweden, August 16, 1960, General Records of the Department of State, Bureau of European Affairs, Office of Soviet Union Affairs, Subject Files, 1957–1963, Box 6, 12 Satellite and Missile Programs, RG 59, NARA.

46. Khrushchev was referring to the U.S. Tiros weather satellite launched in August.
47. Foy D. Kohler, Bureau of European Affairs, to Phillip J. Farley, SAMOS, July 18, 1960, General Records of the Department of State, Bureau of European Affairs, Office of Soviet Union Affairs, Subject Files 1957–63, Box 6, 12 Satellite and Missile Programs, RG 59, NARA.

CHAPTER 6. THE NATIONAL RECONNAISSANCE OFFICE

1. This essay is based primarily on classified records maintained by the NRO, CIA, and Air Force, as well as a number of classified interviews of the major participants. It was not possible to have these records reviewed and declassified or sanitized for inclusion in this essay. All unclassified and secondary sources are clearly identified. Where still-classified sources are used, the phrase "Contained in classified document" is placed in the note to indicate the citation of a classified source. The story of the origins and early years of the NRO in this essay remains valid, even with these limitations.
2. Unlike the National Security Agency (NSA), the NRO does not analyze its own intelligence "take." Analysis is done by the Central Intelligence Agency (CIA), the National Security Agency (NSA), the Defense Intelligence Agency (DIA), the Department of State, and the Central Imagery Office (CIO). The military services also individually have the capability to exploit overhead reconnaissance. The U.S. overhead reconnaissance systems also viewed other denied areas, such as the People's Republic of China.
3. Under Public Law 110, June 20, 1949, Congress gave the Director of Central Intelligence the authority to use federal money without the use of vouchers. No other government agency had this authority.
4. This unique arrangement set the precedent for the NRO's later focus on system performance goals rather than technical specifications in its contract negotiations regarding reconnaissance satellites.
5. Peter Gross, *Gentleman Spy: The Life of Allen Dulles* (New York: Houghton Mifflin, 1994), 469–70. See also Christopher M. Andrew, *For the President's Eyes Only: Secret Intelligence and the American Presidency from Washington to Bush* (New York: HarperCollins, 1995), 221–24.
6. With Killian's support, the CIA funded a study for the design of a high-speed aircraft with a small radar cross-section. This eventually led to Project OXCART (the predecessor to the SR-71 Blackbird), the design and development of the world's fastest and highest flying aircraft.

7. Contained in classified document.
8. Bissell, as early as November 1955, had suggested that some type of formal agreement was needed to define the responsibilities of the CIA and the Air Force. Since the U-2 program had run so smoothly without a formal agreement, nothing more was done. Contained in classified document.
9. In 1958 President Eisenhower had merged the Intelligence Advisory Committee (IAC) with the U.S. Communications Intelligence Board (USCIB) to form the U.S. Intelligence Board (USIB). Contained in classified document.
10. See William M. Leary, *The Central Intelligence Agency: History and Documents* (Tuscaloosa: University of Alabama Press, 1984), 86–87, for a review of the issues.
11. Prior to this reorganization, the CIA's scientific and technical intelligence operations were scattered among several offices. The reconnaissance program, under the Development Projects Division, was in the Directorate of Plans (DP); the Office of Scientific Intelligence (OSI) which conducted basic research, was in the Directorate for Intelligence (DI); the Technical Services Division (TSD), which engaged in research and development to provide operational support for clandestine operations, was also part of the DP; Staff D, which ran electronic intercept operations, resided in the DP as well. The new DDR, Dr. Herbert Scoville Jr., had little authority over the overall program, as the Directorate of Intelligence refused to relinquish OSI, and the Directorate of Plans would not give up its TSD. Contained in classified document.
12. The Air Force Space Systems Division became involved in both Air Force satellite and missile programs. Today it is Air Force Space and Missile Systems.
13. Charyk held the post of Director, National Reconnaissance Office (DNRO), for more than a year until he resigned to head the Communications Satellite Corporation (COMSAT) in early 1963.
14. The NRO was staffed by people from the CIA, Air Force, and Navy. In addition, the NRO had a number of people from the Army, Defense Intelligence Agency, and increasingly from NSA. Contained in classified document.
15. Wheelon was determined to enlarge the CIA's role in overhead reconnaissance. He established a Foreign Missile and Space Analysis Center (FMSAC) and hired Carl E. Duckett of the U.S. Army's Redstone Arsenal to head it. Wheelon also enlarged the Directorate of Research and renamed it the Directorate for Science and Technology. Contained in classified document.
16. The Air Force had successfully pressed for, and finally obtained, responsibility for the U-2 overflight program just prior to the Cuban missile crisis and was demanding a fighter version of the supersonic OXCART. See Leary, *Central Intelligence Agency,* 87. Contained in classified document.
17. In addition to personal differences, McMillan and the Air Force saw Wheelon as a major threat to their reconnaissance program. Wheelon was building a technological empire at DS&T and moving into developmental engineering—an area in which the Air Force believed it had exclusive control. Wheelon further alarmed Air Force officials when he persuaded McCone to establish a separate engineering pay scale that enabled the agency to hire top engineers from private

industry. The Air Force simply could not compete. Contained in classified document.

18. McCone established NIPE in 1963 to assist him in running the Intelligence Community.
19. Leary, *Central Intelligence Agency,* 87.
20. The satellites also made possible the monitoring and verification provisions of the Soviet-U.S. arms limitations and nuclear test-ban treaties. Contained in classified document.
21. From its beginning, the NRO also offered support to U.S. military forces by providing targeting and mapping information needed to plan and execute strategic war plans.

CHAPTER 7. ZENIT

Note: This chapter is based on a research paper first presented at the 63rd annual meeting of the Society for Military History (SMH), Arlington, Virginia, on April 19, 1996, and at the Annual Convention of the American Society for Photogrammetry and Remote Sensing (ASPRS), Baltimore, Maryland, on April 22, 1996.

1. Anatoliy Shiriaev and Valeriy Baberdin, "Before the First Leap into Space," *Krasnaya Zvezda,* April 27, 1996, 5 (in Russian).
2. Asif Siddiqi, "The Road to Object D: Early Satellite Studies in the Soviet Union, 1947–1957," report at the 28th National Convention of American Association for the Advancement of Slavic Studies, Boston, November 14–17, 1996, 4.
3. Boris Rauschenbach et al., *Materials on the History of Vostok Spaceship* (Moscow: Nauka, 1991), 209 (in Russian).
4. S. P. Korolev Space Corporation Energiya, *S. P. Korolev Space Corporation Energiya: To 50th Anniversary* (Moscow: RKK Energiya, 1996), 86, 105 (in Russian).
5. Shiriaev and Baberdin, "Before the First Leap into Space," 5.
6. Korolev Space Corp., *Korolev Space Corporation Energiya,* 87.
7. Mstislav Keldysh, ed., *Creative Legacy of Academician Sergei Pavlovich Korolev* (Moscow: Nauka, 1980), 373 (in Russian).
8. Korolev Space Corp., *Korolev Space Corporation Energiya;* Yuri Frumkin, "The First Reconnaissance Satellite," *Aviatsiya i Kosmonavtika,* no. 3 (1993): 41 (in Russian).
9. Keldysh, ed., *Creative Legacy of Korolev,* 90.
10. Korolev Space Corp., *Korolev Space Corporation Energiya,* 99.
11. Ibid., 98, 108.
12. Keldysh, *Creative Legacy of Korolev,* 373.
13. Korolev Space Corporation Energiya, *Korolev Space Corporation Energiya,* 108.
14. Yuri Frumkin, "Development of the First Soviet Photo-Reconnaissance Satellite Zenit," *Priroda,* no. 4 (1993): 41 (in Russian).
15. Korolev Space Corp., *Korolev Space Corporation Energiya,* 99.
16. Boris Rauschenbach, "The History of the First Stage of Spacecraft Control Systems Development in the USSR," report at the 44th International Astronautical Federation Congress, Graz, Austria, October 16–22, 1993, IAA.2.2-93-67, p. 2.

17. Frumkin, "Development of Zenit," 18.
18. Ibid.
19. Ibid.
20. Korolev Space Corp., *Korolev Space Corporation Energiya;* V. Agapov, "Launches of Zenit-2 Spacecraft," *Novosti Kosmonavtiki* 6, no. 10/125 (1996): 66 (in Russian).
21. Dmitriy Kozlov, ed., *Development of Automated Spacecraft* (Moscow: Mashinostroenie, 1996), 218 (in Russian).
22. Frumkin, "Development of Zenit," 16.
23. Korolev Space Corp., *Korolev Space Corporation Energiya,* 101; Frumkin, "Development of Zenit," 17.
24. Nikolay Dolgopolov, "From Above, We Can See a Ball on a Field in Arizona: An Interview with Victor Nekrasov," *Komsomolskaya Pravda,* July 3, 1993, 4 (in Russian).
25. Kozlov, *Development of Automated Spacecraft,* 17.
26. Maxim Tarasenko, *Military Aspects of the Soviet Cosmonautics* (Moscow: Nikol Press, 1992), 51 (in Russian).
27. Kozlov, ed., *Development of Automated Spacecraft,* 14.
28. Agapov, "Launches of Zenit-2 Spacecraft," 66.
29. Korolev Space Corp., *Korolev Space Corporation Energiya,* 100.
30. Agapov, "Launches of Zenit-2 Spacecraft," 67.
31. Ibid., 76.
32. Korolev Space Corp., *Korolev Space Corporation Energiya,* 100.
33. Agapov, "Launches of Zenit-2 Spacecraft," 42.
34. Korolev Space Corp., *Korolev Space Corporation Energiya,* 101.
35. "Beginning of the Space Era," in *Roads to Space,* vol. 3 (Moscow: RNITsKD, 1994), 264 (in Russian).
36. Ibid., 265
37. Ibid., 272.
38. Igor Sergeiev, ed., *Chronicle of the Main Events in the History of the Strategic Rocket Forces* (Moscow: TsIPK, 1994), 17 (in Russian).
39. Frumkin, "The First Reconnaissance Satellite," 15.
40. Nicholas Johnson and David Rodvold, *Europe and Asia in Space, 1993–1994* (Colorado Springs, Colo.: Kaman Sciences Corporation, 1995), 339.
41. Tarasenko, *Military Aspects of the Soviet Cosmonautics,* 126.
42. Agapov, "Launches of Zenit-2 Spacecraft," 77.
43. Kozlov, *Development of Automated Spacecraft,* 17.
44. *Rocket/Space Industry of Russia,* catalog of enterprises, Russian Space Agency, 1996, 178.
45. Joseph Heyman, *Spacecraft Tables, 1957–1990* (San Diego: Univelt, 1991), 109.
46. Agapov, "Launches of Zenit-2 Spacecraft," 69–77.

CHAPTER 8. CORONA AND THE U.S. PRESIDENTS

1. General Goodpaster has served in a number of military and government positions including Commander-in-Chief of U.S. European Command and Supreme Allied

Commander of Europe. Goodpaster was also responsible for helping President Nixon organize his foreign policy and international security affairs staffs. After retiring in 1974, Goodpaster became a Senior Fellow with the Woodrow Wilson International Center for Scholars at the Smithsonian Institution and an Assistant to Vice-President Rockefeller on the Commission for the Organization of the Government for the Conduct of Foreign Policy. In 1977, he returned to active duty as the 51st Superintendent of the U.S. Military Academy. Currently, General Goodpaster is serving as the chairman of the Atlantic Council of the United States.

2. General Smith joined the Kennedy White House in July 1961. In all, Smith served 35 years in the U.S. Air Force from 1948 through 1983. His last two Air Force assignments were as Chief of Staff of the Supreme Headquarters Allied Powers in Europe (SHAPE) and as Deputy Commander of the U.S.-European Command. Smith also served as military assistant to two secretaries of the Air Force and as assistant to three chairmen of the Joint Chiefs of Staff. When Smith retired from military service in 1983, he accepted an appointment as a Fellow of the Woodrow Wilson International Center for Scholars at the Smithsonian Institution. He then became president of the Institute for Defense Analysis, a government-funded research and development center. He retired from that position in 1991, but continues to serve as president emeritus.
3. Dino Brugioni—the manager of the National Photographic Interpretation Center during the CORONA program (see chapter 11)—has noted that despite Eisenhower and Kennedy's political differences, "a close bond developed between General Eisenhower and President Kennedy" after the 1960 presidential election thanks, in part, to CORONA. For instance, Brugioni stated at the CORONA conference: "After President Kennedy was elected, General Eisenhower sat in on several briefings at which Arthur Lundahl and Richard Bissell briefly taught Kennedy the finer points of intelligence and the value of satellite imagery in the decisionmaking process. Many times during his administration, including the Cuban Missile Crisis, Kennedy was in touch with General Eisenhower, and so was President Johnson, for that matter."
4. Comments made by David Doyle—an imagery analyst and manager at NPIC during the CORONA era (see chapter 11)—reveal how differently certain presidents and advisors interpreted photos related to Vietnam. Doyle suggested at the CORONA symposium, "I don't think that Eisenhower, after looking at the imagery that we provided in 1958 and 1959, would have ever gotten involved in Vietnam. After he saw CORONA photos of how the Vietnamese were beginning to penetrate Laos, he told us about the difficulties of fighting a war under such circumstances. He compared it to the way Tito outfoxed the Germans during World War II."
5. Richard Helms began his intelligence career in the Office of Strategic Services (OSS) in 1943. During his service, he worked in Washington, London, Paris, and Berlin, among other places. After the war, he continued in intelligence work, first by joining the OSS's successor organization, and then the CIA. After holding various positions in the clandestine service, Helms ultimately became U.S. ambassador to Iran in 1973. Four years later, he retired from government service

and became president of the Safeer Company in Washington D.C., an international business consulting firm.

CHAPTER 9. THE ORIGIN AND EVOLUTION OF THE CORONA SYSTEM

1. McMahon first joined the CIA in 1951. After completing an overseas assignment and basic training in the U.S. Army, he returned to his regular assignment at the agency, joined its reconnaissance program, and eventually rose to the position of Deputy Director of the Satellite Office. From 1972 through 1982, he served as the CIA's Deputy Director of Operations, Deputy Director of Intelligence, and Executive Director of CIA (a position which made him responsible for the day-to-day management of the agency). McMahon also acted as Deputy Director of Central Intelligence from 1982 until his retirement in March 1986. In September 1986, McMahon joined the Lockheed Missiles and Space Company and served as its executive vice-president and corporate vice-president until August 1988. At that point, he was elected president of the Lockheed Missiles and Space Group and Company. He remained in those positions until his retirement from Lockheed in December 1994.
2. Allen also served as director of the National Security Agency, Commander of Air Force Systems Command, and Chief of Staff of the Air Force. He was the individual who came up with the concept for the CORONA movie, "A Point in Time," and the person who proposed that the Smithsonian Institution preserve some of CORONA's hardware. After his retirement from the Air Force, Allen became director of the Jet Propulsion Laboratory. From 1995–96, he served as a member of the Aspin Commission (later renamed the Brown Commission), which examined the role and capabilites of the U.S. Intelligence Community. Allen is currently chairman of the board of Draper Laboratories and a member of the President's Foreign Intelligence Advisory Board.
3. Plummer served as the Secretary of the Air Force during the Nixon administration. That assignment included responsibilities as the director of the National Reconnaissance Office. After retiring from government service, Plummer returned to Lockheed as its executive vice-president. He then served as chairman of the board of trustees of the Aerospace Corporation until he retired in 1992.
4. Garwin had a distinguished career with the IBM Corporation. At IBM, he made contributions to the design of nuclear weapons, instruments and electronics for research in nuclear and low-temperature physics, computer systems, and military technology. He is currently an IBM Fellow Emeritus at the Thomas J. Watson Research Center in New York, an Adjunct Research Fellow in the Kennedy School of Government at Harvard University, and an Adjunct Professor of Physics at Columbia University. In March 1996, Garwin received the R. V. Jones Intelligence Award of the National Foreign Intelligence Community.
5. Some of the other Land Panel members included Harvard physicist Edward Purcell and James G. Baker, designer of high-performance optics.

6. Levison retired from Itek in 1974 as the senior vice-president for operations. He continued to act as a consultant for Itek during his retirement, in addition to serving a number of other clients including Lockheed Missiles and Space Company and Mark Systems.

   Incidentally, Levison at the CORONA symposium said that he "never got to see one of those [CORONA] launches because Jim Plummer would never let me go down there for security reasons. Plummer was afraid someone would recognize me."
7. Levison noted at the symposium that "from the very beginning, we knew that panoramic cameras, although rich in intelligence, had distortions that made it hard to use for precise measurements. In fact, my good friend Amrom Katz used to say that photogrammetrics had to do with the character *of* the terrain, but intelligence had to do with the characters *on* the terrain."

CHAPTER 10. CORONA AND THE REVOLUTION IN MAPMAKING

1. Gifford has held several senior management positions at the CIA, including posts in the Office of Development and Engineering. She has also served as NPIC's director of the National Exploitation Laboratory and as deputy director of the Office of Communications. In 1994, she became associate deputy director for administration.
2. Mahoney was instrumental in developing and applying advanced triangulation techniques utilizing satellite imagery. When the Defense Mapping Agency (DMA) was formed in 1972, he assumed the position of the DMA Director of Science and Technology. Mahoney also served as the scientific advisor to the DMA Director of Special Program Office for Exploitation and Modernization. He retired in 1986.

   After reviewing an edited version of the transcript of the May 1995 CORONA conference, Dr. Mahoney supplemented his comments with a paper that further details many of the points he touched on during the meeting. Consequently, his material in this chapter is a combination of both his submitted essay and his original comments.
3. The geoid is the imaginary surface within or around the earth that is everywhere normal to the direction of gravity and coincides with the mean sea levels in the oceans. It is the fundamental surface reference for surveying and mapping.
4. Mahoney at the CORONA symposium noted: "The KH-5 ARGON camera was being developed at the same time that improvements were taking place on the KH-4 configuration. It was a 3-inch dedicated mapping camera flown for a short time. One of the things that the KH-5 demonstrated during its short lifetime was that a 3-inch focal length frame camera could achieve an image pointing accuracy of 30 meters in the horizontal and 50 meters in the vertical. With that information, along with orbit determination, MC&G reduced past target location inaccuracies from a maximum error of two or three miles, down to within about a 1000 foot Circular Error—90 percent (CE), and a vertical error of 300 feet (linear

error—90 percent) on the WGS. It was a fantastic improvement and had an immediate impact on our ability to target enemy strategic weapons. However, the eventual MC&G workhorses ended up being the KH-4A and KH-4B systems that flew with 1.5- and 3-inch focal lengths, and reseau calibrated frame cameras. They flew until the end of the CORONA program. We accomplished all of this without compromising the intelligence function." He also noted that "we primarily used the stellar/index camera to determine attitude. If we caught a good star field on film, we could pin the attitude down to within three to five arc seconds. On the ground, this was equivalent to eight to fifteen feet in ground positioning. One could take this error into account during the calculation process and estimate the final effect it would have on the ground point positioning error."

5. Mahoney stated at the symposium: "Their efforts allowed geodetic targeting capability to improve to CE 90 percent = 450 feet, and LE 90 percent = 300 feet. By 1970, the MC&G community could achieve the following accuracy anywhere in the Soviet bloc to meet SAC (Strategic Air Command) SHOELACE ICBM targeting objectives with respect to WGS: CE 90 percent = 1000 feet (launch to target) and CEP = 600 feet (impact error)."
6. To accomplish this task, the DoD screened hundreds of its employees and sent the most qualified to universities throughout the country so that they could receive training in geodesy, orbital sciences, computer science, photographic science, photogrammetry, and other specialities. The training lasted from one to two years.
7. An aggressive MC&G R&D program spawned the development of software, analytical plotters and comparators, carto compilation equipment, high-resolution optical rectifiers, photo processors, technical management enlargement software, panoramic imagery and associated geodetic/gravity data, and other developments.
8. Daugherty was one of the key individuals involved in establishing geodetic and geophysical support for ballistic missile operations. He has held several high-level positions at the DMA including deputy director, which is the DMA's senior civilian position. Daugherty retired in March 1995.
9. Elaine Gifford also stated at the symposium: "I had to chuckle when I heard Ken say he went to look up what a geodesist was when he first joined the team because I flashed back to when I was bent over a dictionary trying to figure out what a 'photogrammetrist' was; I wasn't at all sure what photogrammetry encompassed. But I think these two stories show that we were paving new ground during those days. Although we didn't have specific job titles, we did have the sense of a great mission and what we could do with CORONA. The CORONA pioneers and innovators were creative and came up with some wonderful ideas to get the most out of the system."
10. William Mahoney at the symposium noted, "By 1972, we had built up a cadre of almost 100 to 200 people who had received advanced college training in photogrammetry, computer science, and other mapping related disciplines, and we used them in any position we needed them in."
11. Lowell Starr, a fellow mapping expert, noted at the CORONA symposium: "It's interesting that you bring up how the Soviets obtained maps of the United States because we used to have a map information office over at 19th and E Street here

in [Washington] D.C.—when the USGS [U.S. Geological Survey] was in the old GSA [Government Services Administration] building—and there was this little lady from the Russian embassy who would come in every month and get all of the new publications. It was so routine that after a while the people working at the desk would anticipate her arrival and would stack up all of the new publications for her. Then, she'd just walk in, count them, pay for them, and leave. This went on forever."

12. The Global Positioning System is a network of DoD-operated earth-orbiting satellites that provides navigational positioning data to military and civilian users.
13. From 1986 until 1991, Starr served as chief of the National Mapping Program. As chief, he was responsible for the production of multipurpose maps, the development of a digital database of geographic and cartographic information, and the investigation of the cartographic application of remote sensing data. After his retirement from government service, Starr took a position as the technical advisor for the International Federal Systems of the Intergraph Corporation.
14. However, at the symposium, Starr suggested that overall "there was a tremendous amount of cooperation and synergism between the various Intelligence Community organizations, including the DIA, the ACIC, the Army Map group, and the DMA."

CHAPTER 11. EXPLOITING CORONA IMAGERY

1. Huffstutler served with the CIA for 35 years. During the first 25, he held senior management posts in the Office of Weapons Analysis, the Office of Strategic Research, and the Office of Soviet Analysis. In the mid-1980s, he became the director of the National Photographic Interpretation Center (NPIC), and from 1988 to 1991, he acted as the CIA's deputy director for administration. Huffstutler completed his career with the CIA in 1994, after two years of service as the agency's executive director. He currently works for the Aegis Research Corporation.
2. David Doyle, another former imagery analyst and manager at NPIC during the CORONA era, added at the CORONA symposium: "The questions we were being asked in the early days were very basic. 'Where are the airfields?' Not 'What kind of airplanes are on them,' as much as just 'Where are the installations and where are the missile sites?'"
3. Kerr joined the CIA in 1960 shortly after the CORONA program began. He originally worked with the team of analysts who were responsible for writing CORONA mission summaries. Kerr later became the executive officer for the Intelligence Community Staff and vice chairman of the Committee on Imagery Requirements and Exploitation (COMIREX). From 1976 to 1982, he served as the director or deputy director of several offices responsible for worldwide political analysis and current intelligence. He became associate deputy director for intelligence in 1982. Four years later, he received an appointment as the CIA's deputy director for intelligence, and then in 1989 became the deputy DCI. Kerr retired from the CIA in 1992.

4. Brugioni is considered "one of the founding fathers of NPIC." During the CORONA era, he was responsible for writing the summaries that NPIC's director Arthur Lundahl used to brief presidents, Congress, the cabinet officers, and the Joint Chiefs of Staff. Brugioni has written extensively on the application of imagery to intelligence and is the author of a detailed analysis of aerial photographs of the Nazi death camp Auschwitz-Birkenau. After his retirement from the CIA, he wrote *Eyeball to Eyeball: The Inside Story of the Cuban Missile Crisis.* He continues to write and lecture on subjects related to overhead imagery.
5. David Doyle, at the CORONA symposium, added: "The Chinese, however, were not quite as helpful to us. They did not build fences around their installations. Their missile sites were very difficult to detect."

CONTRIBUTORS

Dwayne A. Day is a military space policy analyst and historian in Virginia. He was the 1996–97 Guggenheim Fellow at the National Air and Space Museum and a NASA Space Grant Fellow at the Space Policy Institute of George Washington University's Elliott School of International Affairs. He is the author of numerous articles on American military and civilian space programs, an assistant editor of *Exploring the Unknown: Selected Documents in the History of the U.S. Civil Space Program,* vols. 1 and 2, and editor of several issues of the *Journal of the British Interplanetary Society* on military space history. He is currently working on two books about science and the Air Force and the race to the moon.

Peter A. Gorin is a specialist in political science and a longtime researcher of aerospace history. He graduated from the Moscow University and worked as a political analyst. He continued his research in association with the National Air and Space Museum as a Guggenheim Fellow. Mr. Gorin is a contributor to *Air & Space* and *Quest* magazines, as well as to the public TV (WGBH) presentation "Spy in the Sky" (1996).

Gerald Haines is the historian for the director of Central Intelligence. During 1996 and 1997, he was the historian for the National Reconnaissance Office. He was previously deputy chief historian, director of Central Intelligence History Staff, and chief of the History Branch at the National Security Agency. He is the author of *Unlocking the Files of the FBI: A Guide to Its Records and Classification System* as well as numerous articles on diplomatic history.

R. Cargill Hall is the historian at the National Reconnaissance Office. He has served as the historian at NASA's Jet Propulsion Laboratory and as chief of the Research Division at the Air Force Historical Research Agency. Hall is the author of *Lunar Impact: A History of Project Ranger,* and editor and contributor to *Lightning Over Bougainville.*

Brian Latell is the former director of the CIA's Center for the Study of Intelligence and the former editor of the CIA journal *Studies in Intelligence.* He is currently a professor at Georgetown University and specializes in the history of Central America.

John M. Logsdon is the director of the Space Policy Institute and the Center for International Science and Technology Policy at the Elliott School of International Affairs, George Washington University where he is professor of political science and international affairs.

Ernest R. May is director of the Charles Warren Center for Studies in American History at Harvard University. He has also been dean of Harvard College, associate dean of the Faculty of Arts and Sciences, director of the Institute on Politics, and chairman of the Department of History. He has written extensively on American foreign policy and the uses of history in decisionmaking.

Jonathan McDowell is a writer and researcher on the technical history of the space program, and has published papers on the topic in *Quest* magazine and the *Journal of the British Interplanetary Society,* as well as writing a monthly column in *Sky and Telescope* magazine. Since 1989 he has distributed a weekly electronic newsletter with details on current space activities, *Jonathan's Space Report* (http://hea-www.harvard.edu/~jcm/space/jsr/jsr.html), which is widely distributed in the aerospace community. Dr. McDowell is also an astrophysicist at the Smithsonian Astrophysical Observatory in Cambridge, Mass., and has published numerous scientific papers on quasars, black holes, and cosmology.

Albert "Bud" Wheelon was the CIA's first deputy director for science and technology, a position he held from 1963 to 1966. He later joined Hughes Aircraft Company as vice-president for engineering and was named executive vice-president for operations in 1986. He became chief executive officer and chairman of the board in 1987. He served as visiting professor of science, technology, and policy at MIT in 1989.

INDEX

Page numbers in bold indicate illustrations